Lecture Notes in Computer Science 13542

More information about this series at https://link.springer.com/bookseries/558

Konstantinos Kamnitsas · Lisa Koch ·
Mobarakol Islam · Ziyue Xu · Jorge Cardoso ·
Qi Dou · Nicola Rieke · Sotirios Tsaftaris (Eds.)

Domain Adaptation and Representation Transfer

4th MICCAI Workshop, DART 2022
Held in Conjunction with MICCAI 2022
Singapore, September 22, 2022
Proceedings

Editors
Konstantinos Kamnitsas
University of Oxford
Oxford, UK

Lisa Koch
University of Tübingen
Tübingen, Germany

Mobarakol Islam
Imperial College London
London, UK

Ziyue Xu
Nvidia Corporation
Santa Clara, CA, USA

Jorge Cardoso
King's College London
London, UK

Qi Dou
Chinese University of Hong Kong
Hong Kong, Hong Kong

Nicola Rieke
Nvidia GmbH
Munich, Bayern, Germany

Sotirios Tsaftaris
University of Edinburgh
Edinburgh, UK

ISSN 0302-9743 ISSN 1611-3349 (electronic)
Lecture Notes in Computer Science
ISBN 978-3-031-16851-2 ISBN 978-3-031-16852-9 (eBook)
https://doi.org/10.1007/978-3-031-16852-9

This Springer imprint is published by the registered company Springer Nature Switzerland AG
The registered company address is: Gewerbestrasse 11, 6330 Cham, Switzerland

Preface

Computer vision and medical imaging have been revolutionized by the introduction of advanced machine learning and deep learning methodologies. Recent approaches have shown unprecedented performance gains in tasks such as segmentation, classification, detection, and registration. Although these results (obtained mainly on public datasets) represent important milestones for the Medical Image Computing and Computer Assisted Intervention (MICCAI) community, most methods lack generalization capabilities when presented with previously unseen situations (corner cases) or different input data domains. This limits clinical applicability of these innovative approaches and therefore diminishes their impact. Transfer learning, representation learning, and domain adaptation techniques have been used to tackle problems such as model training using small datasets while obtaining generalizable representations; performing domain adaptation via few-shot learning; obtaining interpretable representations that are understood by humans; and leveraging knowledge learned from a particular domain to solve problems in another.

The fourth MICCAI workshop on Domain Adaptation and Representation Transfer (DART 2022) aimed at creating a discussion forum to compare, evaluate, and discuss methodological advancements and ideas that can improve the applicability of machine learning (ML)/deep learning (DL) approaches to clinical settings by making them robust and consistent across different domains.

During the fourth edition of DART, 25 papers were submitted for consideration and, after peer review, 13 full papers were accepted for presentation. Each paper was rigorously reviewed by at least three reviewers in a double-blind review process. The papers were automatically assigned to reviewers, taking into account and avoiding potential conflicts of interest and recent work collaborations between peers. Reviewers were selected from among the most prominent experts in the field from all over the world. Once the reviews were obtained, the area chairs formulated final decisions over acceptance or rejection of each manuscript. These decisions were always taken according to the reviews and were unappealable.

Additionally, the workshop organization committee granted the Best Paper Award to the best submission presented at DART 2022. The Best Paper Award was assigned as a result of a secret voting procedure where each member of the committee indicated two papers worthy of consideration for the award. The paper collecting the majority of votes was then chosen by the committee.

We believe that the paper selection process implemented during DART 2022, as well as the quality of the submissions, resulted in scientifically validated and interesting contributions to the MICCAI community and, in particular, to researchers working on domain adaptation and representation transfer.

We would therefore like to thank the authors for their contributions and the reviewers for their dedication and professionality in delivering expert opinions about the submissions.

August 2022

M. Jorge Cardoso
Qi Dou
Mobarakol Islam
Konstantinos Kamnitsas
Lisa Koch
Nicola Rieke
Sotirios Tsaftaris
Ziyue Xu

Organization

Organizers

Cardoso, M. Jorge	King's College London, UK
Dou, Qi	The Chinese University of Hong Kong, China
Islam, Mobarakol	Imperial College London, UK
Kamnitsas, Konstantinos	University of Oxford, UK
Koch, Lisa	University of Tuebingen, Germany
Rieke, Nicola	NVIDIA, Germany
Tsaftaris, Sotirios	University of Edinburgh, UK
Xu, Ziyue	NVIDIA, USA

Program Committee

Alsharid, Mohammad	University of Oxford, UK
Bagci, Ulas	Northwestern University, USA
Bai, Wenjia	Imperial College London, UK
Banerjee, Abhirup	University of Oxford, UK
Bredell, Gustav	ETH Zurich, Switzerland
Chaitanya, Krishna	ETH Zurich, Switzerland
Chen, Cheng	The Chinese University of Hong Kong, China
Dinsdale, Nicola	University of Oxford, UK
Feng, Ruibin	Stanford University, USA
Gao, Mingchen	University at Buffalo, SUNY, USA
Guan, Hao	University of North Carolina at Chapel Hill, USA
Haghighi, Fatemeh	Arizona State University, USA
Hosseinzadeh Taher, Mohammad Reza	Arizona State University, USA
Jiménez-Sánchez, Amelia	IT University of Copenhagen, Denmark
Karani, Neerav	MIT, USA
Laves, Max-Heinrich	Philips Research, Germany
Li, Zeju	Imperial College London, UK
Liu, Xiao	University of Edinburgh, UK
Liu, Xinyu	City University of Hong Kong, China
Mahapatra, Dwarikanath	Inception Institute of Artificial Intelligence, UAE
Manakov, Ilja	ImFusion, Germany
Meng, Qingjie	Imperial College London, UK
Menten, Martin	Technische Universität München, Germany

Namburete, Ana	University of Oxford, UK
Nicolson, Angus	University of Oxford, UK
Ouyang, Cheng	Imperial College London, UK
Paschali, Magdalini	Stanford University, USA
Prevost, Raphael	ImFusion, Germany
Saha, Pramit	University of Oxford, UK
Sanchez, Pedro	University of Edinburgh, UK
Sarhan, Hasan	Johnson & Johnson, Germany
Seenivasan, Lalithkumar	National University of Singapore, Singapore
Sudre, Carole	University College London, UK
Thermos, Spyridon	University of Edinburgh, UK
Valanarasu, Jeya Maria Jose	Johns Hopkins University, USA
Xia, Yong	Northwestern Polytechnical University, China
Xu, Yan	Beihang University, China
Xu, Mengya	National University of Singapore, Singapore
Yuan, Yixuan	City University of Hong Kong
Zakazov, Ivan	Philips Research, Russia
Zhang, Kangning	University of Oxford, UK
Zhang, Yue	Siemens Healthineers, USA
Zimmerer, David	German Cancer Research Center (DKFZ), Germany

Contents

Detecting Melanoma Fairly: Skin Tone Detection and Debiasing for Skin Lesion Classification

Peter J. Bevan[1(✉)] and Amir Atapour-Abarghouei[2]

[1] School of Computing, Newcastle University, Newcastle upon Tyne, UK
peterbevan@hotmail.co.uk
[2] Department of Computer Science, Durham University, Durham, UK

Abstract. Convolutional Neural Networks have demonstrated human-level performance in the classification of melanoma and other skin lesions, but evident performance disparities between differing skin tones should be addressed before widespread deployment. In this work, we propose an efficient yet effective algorithm for automatically labelling the skin tone of lesion images, and use this to annotate the benchmark ISIC dataset. We subsequently use these automated labels as the target for two leading bias 'unlearning' techniques towards mitigating skin tone bias. Our experimental results provide evidence that our skin tone detection algorithm outperforms existing solutions and that 'unlearning' skin tone may improve generalisation and can reduce the performance disparity between melanoma detection in lighter and darker skin tones.

1 Introduction

Convolutional Neural Networks (CNN) have demonstrated impressive performance on a variety of medical imaging tasks, one such being the classification of skin lesion images [2,3,10]. However, there are also many potential pitfalls that must be identified and mitigated before widespread deployment to prevent the replication of mistakes and systematic issues on a massive scale. For example, an issue that is commonly raised in the existing literature is skin tone bias in lesion classification tasks. Groh et al. [9] provide a compiled dataset of clinical lesions with human annotated Fitzpatrick skin type [7] labels, and show that CNNs perform best at classifying skin types similar to the skin types in the training data used. We use the skin type labels in this dataset as the target for supervised debiasing methods to evaluate the effectiveness of these methods at helping melanoma classification models generalise to unseen skin types.

Once we have evaluated the effectiveness of the debiasing methods using human labelled skin tone labels, we look to automate the pipeline further, since human annotated labels are expensive and impractical to gather in practice. We

---

Supplementary Information The online version contains supplementary material available at https://doi.org/10.1007/978-3-031-16852-9_1.

K. Kamnitsas et al. (Eds.): DART 2022, LNCS 13542, pp. 1–11, 2022.
https://doi.org/10.1007/978-3-031-16852-9_1

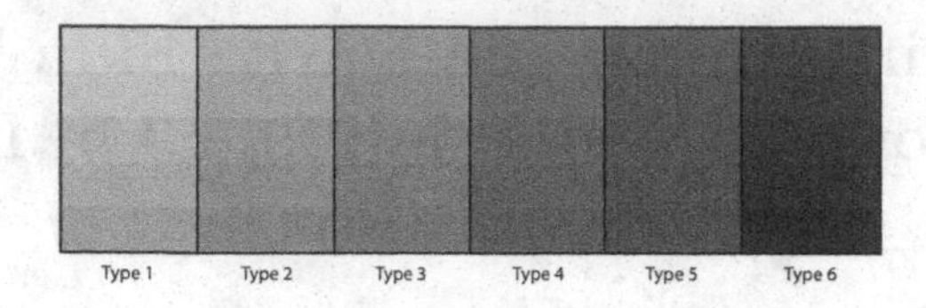

Fig. 1. Visualisation of the Fitzpatrick 6 point scale [7], widely accepted as the gold standard amongst dermatologists [4].

use a novel variation on the skin tone labelling algorithm presented in [16] to annotate the ISIC data and subsequently use these generated labels as the target for a debiasing head, towards creating a fully automated solution to improving the generalisation of models to images of individuals from differing ethnic origins.

In summary, our primary contributions towards the discussed issues are:

- *Skin tone detection* - We propose an effective skin tone detection algorithm inspired by [16] (Sect. 4.2), the results of which can be used as labels for skin tone bias removal.
- *Skin tone debiasing* - We assess the effectiveness of leading debiasing methods [1,15] for skin tone bias removal in melanoma classification, and implement these using automated labels as the target for debiasing (Sects. 4.1 and 4.3).

Code is available at https://github.com/pbevan1/Detecting-Melanoma-Fairly.

2 Related Work

Groh et al. [9] illustrate that CNNs perform better at classifying images with similar skin tones to those the model was trained on. Performance is, therefore, likely to be poor for patients with darker skin tones when the training data is predominantly images of light-skinned patients, which is the case with many of the current commonly-used dermoscopic training datasets such as the ISIC archive data [5,19]. While melanoma incidence is much lower among the black population (1.0 per 100,000 compared to 23.5 per 100,000 for whites), 10-year melanoma-specific survival is lower for black patients (73%) than white patients (88%) or other races (85%) [6], and so it is of heightened importance to classify lesions in patients of colour correctly.

One way to ensure a more even classification performance across skin tones is to re-balance the training data by collecting more high-quality images of lesions on skin of colour, but the low incidence of melanoma in darker skin means this could be a slow process over many years. During the time that unbalanced data continues to be an issue, a robust automated method for removing skin tone bias from the model pipeline could potentially help models to operate with increased fairness across skin tones.

3 Methods

3.1 Debiasing Methods

In this work, two leading debiasing techniques within the literature are used, namely 'Learning Not To Learn' (LNTL) [15] and 'Turning a Blind Eye' (TABE) [1]. Both are often referred to as 'unlearning' techniques because of their ability to remove bias from the feature representation of a network by minimising the mutual information between the feature embedding and the unwanted bias. Generic schematics of both 'Learning Not to Learn' and 'Turning a Blind Eye' are shown in Fig. 2.

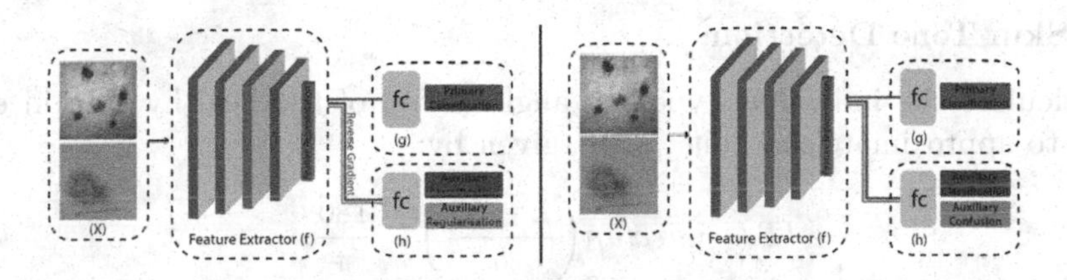

Fig. 2. 'Learning Not to Learn' architecture (left) and 'Turning a Blind Eye' architecture (right). f is implemented as a convolutional architecture such as ResNeXt or EfficientNet in this work. 'fc' denotes a fully connected layer.

Learning Not to Learn. 'Learning Not to Learn' (LNTL) [15] introduces a secondary regularisation loss in combination with a gradient reversal layer [8] to remove a target bias from the feature representation of a CNN during training.

The input image, x, is passed into a CNN feature extractor, f: $x \to \mathbb{R}^K$, where K is the dimension of the embedded feature.

The extracted feature embedding is then passed in parallel into the primary classification head g: $\mathbb{R}^K \to \mathcal{Y}$ and the secondary bias classification head h: $\mathbb{R}^K \to \mathcal{B}$. $\mathcal{Y}$ denotes the set of possible lesion classes and $\mathcal{B}$ denotes the target bias classes.

Formulated as a minimax game, h minimises cross-entropy, learning to classify bias from the extracted features, whilst f maximises cross-entropy, restraining h from predicting the bias, and also minimises negative conditional entropy, reducing the mutual information between the feature representation and the bias. The gradient reversal layer between h and f is used as an additional step to remove information relating to the target bias from the feature representation by multiplying the gradient of the secondary classification loss by a negative scalar during backpropagation, further facilitating the feature extraction network, f, to 'unlearn' the targeted bias, $b(x)$. On completion of training, f extracts a feature embedding absent of bias information, g uses this feature embedding to perform an unbiased primary classification, and the performance of h has deteriorated because of the resulting lack of bias signal in the feature embedding.

Turning a Blind Eye. 'Turning a Blind Eye' (TABE) also removes unwanted bias using a secondary classifier, θ_m, m being the m-th bias to be removed. The TABE secondary classifier identifies bias in the feature representation θ_{repr} by minimising a secondary classification loss, $\mathcal{L}_s$, and also a secondary confusion loss [22], $\mathcal{L}_{\text{conf}}$, which pushes θ_{repr} towards invariance to the identified bias. The losses are minimised in separate steps since they oppose one another: $\mathcal{L}_s$ is minimised alone, followed by the primary classification loss, $\mathcal{L}_p$, together with $\mathcal{L}_{\text{conf}}$. The confusion loss calculates the cross entropy between a uniform distribution and the output predicted bias.

As suggested in [15], TABE can also apply gradient reversal (GR) to the secondary classification loss, and is referred to as 'CLGR' in this work.

3.2 Skin Tone Detection

We calculate the individual typology angle (ITA) of the healthy skin in each image to approximate skin tone [9,16], given by:

$$ITA = arctan\left(\frac{L-50}{b}\right) \times \frac{180}{\pi}, \tag{1}$$

where L and b are obtained by converting RGB pixel values to the CIELAB colour space. We propose a simpler and more efficient method for isolating healthy skin than the segmentation method used in [9,16]. Across all skin tones, lesions and blemishes are mostly darker than the surrounding skin. Consequently, to select a non-diseased patch of skin, we take 8 samples of 20×20 pixels from around the edges of each image and use the sample with the highest ITA value (lightest skin tone) as the estimated skin tone. The idea behind replacing segmentation with this method is to reduce the impact of variable lighting conditions on the skin tone estimation by selecting the lightest sample rather than the entire healthy skin area. This method is also quicker and more efficient than segmentation methods due to its simplicity.

Equation 2 shows the thresholds set out in [9], which are taken from [16] and modified to fit the Fitzpatrick 6 point scale [7] (see Fig. 1). We use these thresholds in our skin tone labelling algorithm.

$$Fitzpatrick(ITA) = \begin{cases} 1 & ITA > 55 \\ 2 & 55 \geq ITA > 41 \\ 3 & 41 \geq ITA > 28 \\ 4 & 28 \geq ITA > 19 \\ 5 & 19 \geq ITA > 10 \\ 6 & 10 \geq ITA \end{cases} \tag{2}$$

We pre-process each image using black-hat morphology to remove hair, preventing dark pixels from hairs skewing the calculation. This hair removal is purely for skin tone detection and the original images are used for training the debiased classification models. It is clear that even with large lesions with hard-to-define borders, our method is highly likely to select a sample of healthy skin (Fig. 3).

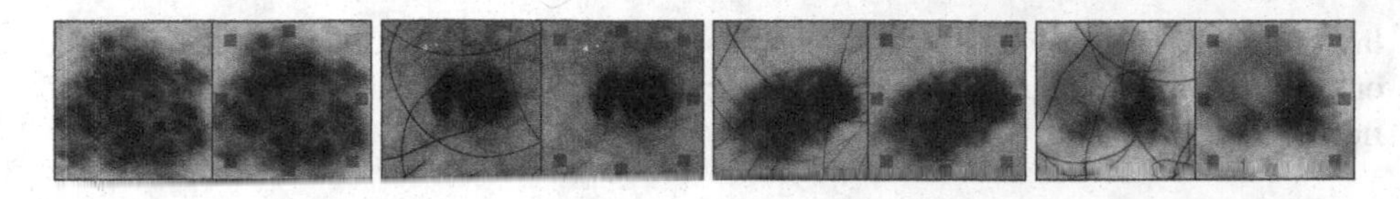

Fig. 3. Left of each pair shows ISIC input images, right of each pair shows the placement of the 20×20 pixel samples on images with hair removed. Green square indicates chosen sample based on lightest calculated tone. This sampling method eliminates the need for segmentation. (Color figure online)

3.3 Data

Training Data. A compilation of clinical skin condition images with human annotated Fitzpatrick skin types [7], called the 'Fitzpatrick17k' dataset [9], is used for training to demonstrate the effectiveness of unlearning for skin tone debiasing, and to evaluate our automated skin tone labelling algorithm. Of the 16,577 images, we focus on the 4,316 of these that are neoplastic (tumorous). These labels are provided by non-dermatologist annotators, so are likely to be imperfect. When attempting dibiasing of ISIC data, a combination of the 2017 and 2020 challenge data [5,19] (35,574 images) is used as training data.

Test Data. The MClass [3] dataset is used to evaluate generalisation and provide a human benchmark. This dataset comprises a set of 100 dermoscopic images and 100 clinical images (*different* lesions), each with 20 malignant and 80 benign lesions. The human benchmark is the classification performance of 157 dermatologists on the images in the dataset. The Interactive Atlas of Dermoscopy [17], and the ASAN datasets [11] were used to further test the robustness of the models. The Atlas dataset has 1,000 lesions, with one dermoscopic and one clinical image per lesion (2,000 total), while the ASAN test dataset has 852 images, all clinical. Whilst the ISIC training data [5,19] is mostly white Western patients, the Atlas seems to have representation from a broad variety of ethnic groups, and ASAN from predominantly South Korean patients, which should allow for a good test of a model's ability to deal with different domain shifts.

3.4 Implementation

PyTorch [18] is used to implement the models. The setup used for experimentation consists of two NVIDIA Titan RTX GPUs in parallel with a combined memory of 48 GB on an Arch Linux system with a 3.30 GHz 10-core Intel CPU and 64 GB of memory. The source code is publicly released to enable reproducibility and further technical analysis.

After experimentation with EfficientNet-B3 [21], ResNet-101 [12], ResNeXt-101 [23], DenseNet [13] and Inception-v3 [20], ResNeXt-101 looked to show the best performance and so was used as the feature extractor in the debiasing experiments. All classification heads are implemented as one fully-connected layer, as in [15]. Stochastic gradient descent (SGD) is used across all models, ensuring comparability and compatibility between the baseline and debiasing networks.

4 Experimental Results

4.1 Fitzpatrick17k Skin Tone Debiasing

A CNN trained using Fitzpatrick [7] types 1 and 2 skin is shown to perform better at classifying skin conditions in types 3 and 4 than types 5 and 6 skin in [9]. We are able to reproduce these findings with our baseline ResNeXt-101 model, trained and tested on the neoplastic subset of the Fitzpatrick17k data. Our objective is to close this gap with the addition of a secondary debiasing head which uses skin type labels as its target. The CLGR configuration proves to be most effective, and is shown in Table 1. The disparity in AUC between the two groups is closed from 0.037 to 0.030, with types 3 and 4 boosted by 1.3% and types 5 and 6 boosted by 2.2%. It is important to note that due to the critical nature of the problem and the significant ramifications of false predictions in real-world applications, even small improvements are highly valuable. This experiment serves as a proof of concept for the mitigation of skin tone bias with unlearning techniques, and gives us precedent to explore this for debiasing the ISIC [5,19] or other similar datasets. Since the ISIC data does not have human annotated skin tone labels, to explore debiasing this dataset we first generate these labels with an automated skin tone labelling algorithm (see Sect. 4.2).

4.2 Automated Skin Tone Labelling Algorithm

To validate the effectiveness of our skin tone labelling algorithm, we re-label the Fitzpatrick17k data and compare these automated labels against the human annotated skin tones to calculate accuracy, with a correct prediction being within ±1 point on the Fitzpatrick scale [9]. Our method achieves 60.61% accuracy, in comparison to the 53.30% accuracy achieved by the algorithm presented in [9], which segments the healthy skin using a YCbCr masking algorithm. The authors of [9] improve their accuracy to 70.38% using empirically selected ITA thresholds, but we decide against using these to label the ISIC data, given that they are optimised to suit only the Fitzpatrick17k data and do not generalise.

Table 1. Improving model generalisation to skin tones different to the training data [9]. All scores are **AUC**. Trained using types 1&2 skin images from the Fitzpatrick17k dataset [9], tested on types 3&4 skin and types 5&6.

Experiment	Types 3&4	Types 5&6
Baseline	0.872	0.835
CLGR	**0.883**	**0.853**

We expect our algorithm to perform better still on the ISIC data [5,19] than the Fitzpatrick17k data [9], since the images are less noisy, meaning the assumption that the lightest patch in the image is healthy skin is less likely to be undermined by artefacts or a lightly coloured background.

Figure 4 shows the distribution of Fitzpatrick skin types in the ISIC training data, labelled by our skin tone detection algorithm. The figure shows a clear imbalance towards lighter skin tones. The relatively high number of type 6 classifications could be due to the labelling algorithm picking up on dark lighting conditions, since upon visual inspection of the dataset, it can be concluded that there is not likely to be this many type 6 skin images in the dataset. This is something that should be explored and improved in future work.

4.3 ISIC Skin Tone Debiasing

The ISIC archive is one of the most popular publicly available melanoma training datasets, but there are no skin tone labels available, so we use our skin tone labelling algorithm to analyse the distribution of skin tones in this data as well as to further test the debiasing methods. We also use these labels as the target for the debiasing heads during training. Although these labels have low accuracy, it has been shown that deep learning is still able to learn, even in cases where labels are noisy [14]. We see a small performance improvement across the board when debiasing with the TABE [1] head, indicating that this model generalises to the test sets better than the baseline (see Table 2), including a 5.3% improvement in AUC on the ASAN test set. Performance on this dataset is of particular interest since these images are known to be from Korean patients and so represent a definitive domain shift in comparison to the predominantly Western ISIC training data. The TABE head also prompts a 14.8% increase in performance on the Atlas clinical test set [17] compared to the baseline, and all debiasing heads show noticeable improvements on the MClass dermoscopic and clinical test sets [3]. Although the origins of the Atlas and MClass clinical data are unknown, these also look to be drawn from significantly different populations to the ISIC data (containing many

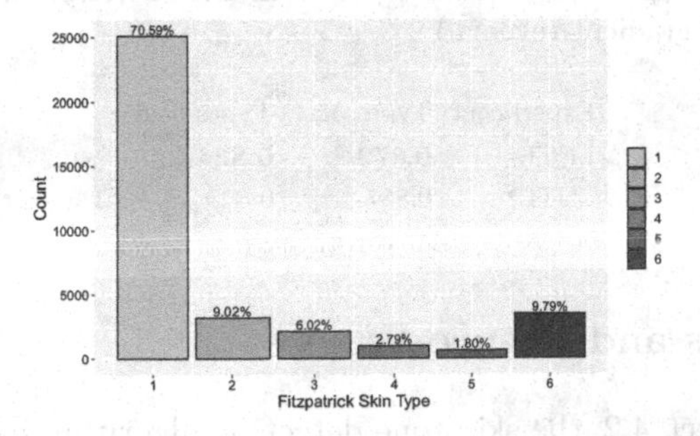

Fig. 4. Distribution of Fitzpatrick skin types in ISIC [5,19] training data, as labelled by our algorithm.

more examples of darker skin tones), so improvements on these test sets could be interpreted as evidence of the mitigation of skin tone bias.

Our models demonstrate superior classification performance compared to the group of dermatologists from [3]. While impressive, this comparison should be taken with a grain of salt, as these dermatologists were classifying solely using images and no other information. A standard clinical encounter with each patient would likely result in better performance than this. Moreover, systems like this are not meant to replace the expertise of a dermatologist at this stage, but to augment and enhance the diagnosis and facilitate easier access to certain patients.

Table 2. Comparison of skin tone debiasing techniques, with AUC used as the primary metric. Models are trained using ISIC 2020 & ISIC 2017 data [5,19].

Experiment	Atlas		Asan	MClass	
	Dermoscopic	Clinical	Clinical	Dermoscopic	Clinical
Dermatologists	—	—	—	0.671	0.769
Baseline	0.819	0.616	0.768	0.853	0.744
LNTL	0.803	0.608	0.765	0.858	0.787
TABE	**0.825**	**0.707**	**0.809**	0.865	**0.859**
CLGR	0.820	0.641	0.740	**0.918**	0.771

4.4 Ablation Studies

TABE [1] with and without gradient reversal has provided impressive results, but ablation of the gradient reversal layer from LNTL [15] led to degraded performance (see Table 3). Deeper secondary heads were experimented with (additional fully-connected layer), but did not have a noticeable impact on performance (see supplementary material).

Table 3. Ablation of gradient reversal layer from LNTL (ResNext101). Asterisk (*) indicates ablation of gradient reversal).

Experiment	Types 3&4	Types 5&6
LNTL	**0.873**	**0.834**
LNTL*	0.867	0.829

5 Limitations and Future Work

As mentioned in Sect. 4.2, the skin tone detection algorithm has a problem with over-classifying type 6 skin which is a key limitation and should be addressed. ITA is an imperfect method for estimating skin tone, given its sensitivity to lighting conditions, and the Fitzpatrick conversion thresholds are tight and may not generalise well. Empirical calibration of these thresholds tailored to the specific data in question may help, as is done in [9].

Further work may collect dermatologist annotated skin tone labels for dermoscopic datasets and evaluate the effectiveness of debiasing techniques using these human labels. These labels would also allow a more robust evaluation of skin tone bias in the ISIC data than we were able to provide.

Although this work provides potential methods for bias mitigation in melanoma detection, we caution against over-reliance on this or similar systems as silver bullet solutions, as this could further lead to the root cause of the problem (imbalance and bias within the data) being overlooked. We encourage a multifaceted approach to solving the problem going forward. Further work may also look to do a deeper analysis into the debiasing methods to confirm that the improved generalisation is a result of mitigation of the targeted bias.

6 Conclusion

This work has provided evidence that the skin tone bias shown in [9] can be at least partially mitigated by using skin tone as the target for a secondary debiasing head. We have also presented an effective variation of Kinyanjui et al.'s skin tone detection algorithm [16], and used this to label ISIC data. We have used these labels to unlearn skin tone when training on ISIC data and demonstrated some improvements in generalisation, especially when using a 'Turning a Blind Eye' [1] debiasing head. Given that current publicly available data in this field is mostly collected in Western countries, generalisation and bias removal tools such as these may be important in ensuring these models can be deployed to less represented locations as soon as possible in a fair and safe manner.

References

1. Alvi, M., Zisserman, A., Nellåker, C.: Turning a blind eye: explicit removal of biases and variation from deep neural network embeddings. In: Leal-Taixé, L., Roth, S. (eds.) ECCV 2018. LNCS, vol. 11129, pp. 556–572. Springer, Cham (2019). https://doi.org/10.1007/978-3-030-11009-3_34
2. Brinker, T.J., et al :A convolutional neural network trained with dermoscopic images performed on par with 145 dermatologists in a clinical melanoma image classification task. European J. Cancer (Oxford, England: 1990), **111** 148–154 (2019)
3. Brinker, T.J., The melanoma classification benchmark, et al.: Comparing artificial intelligence algorithms to 157 German dermatologists. Eur. J. Cancer **111**, 30–37 (2019)
4. Buolamwini, J., Gebru, T.: Gender Shades: Intersectional Accuracy Disparities in Commercial Gender Classification. In Conference on Fairness, Accountability and Transparency, pp. 77–91, PMLR (2018)

5. Codella, N.C.F., et al.: Skin lesion analysis toward melanoma detection: A challenge at the 2017 International symposium on biomedical imaging (ISBI), hosted by the international skin imaging collaboration (ISIC). In 2018 IEEE 15th International Symposium on Biomedical Imaging (ISBI 2018) pp. 168–172 (2018)
6. Collins, K.K., Fields, R.C., Baptiste, D., Liu, Y., Moley, J., Jeffe, D.B.: Racial Differences in Survival after Surgical Treatment for Melanoma. Ann. Surg. Oncol. **18**(10), 2925–2936 (2011)
7. Fitzpatrick, T.B.: The validity and practicality of sun-reactive skin types I through VI. Arch. Dermatol. **124**(6), 869–871 (1988)
8. Ganin, Y., et al.: Domain-adversarial training of neural networks. In: Csurka, Gabriela (ed.) Domain Adaptation in Computer Vision Applications. ACVPR, pp. 189–209. Springer, Cham (2017). https://doi.org/10.1007/978-3-319-58347-1_10
9. Groh, M., et al.: Evaluating Deep Neural Networks Trained on Clinical Images in Dermatology with the Fitzpatrick 17k Dataset. arXiv:2104.09957 [cs], April 2021
10. Haenssle, H.A., Fink, C., Schneiderbauer, R., Toberer, F., Buhl, T., Blum, A., Kalloo, A., et al.: Man against machine: Diagnostic performance of a deep learning convolutional neural network for dermoscopic melanoma recognition in comparison to 58 dermatologists. Ann. Oncol. **29**(8), 1836–1842 (2018)
11. Han, S.S., et al: Classification of the Clinical Images for Benign and Malignant Cutaneous Tumors Using a Deep Learning Algorithm. J. Invest. Dermatol. **138**(7), 1529–1538 (2018)
12. He, K., Zhang, X., Ren, S., Sun, J.: Deep Residual Learning for Image Recognition. In 2016 IEEE Conference on Computer Vision and Pattern Recognition (CVPR), pp. 770–778, Las Vegas, NV, USA, IEEE (2016)
13. Huang, G., Liu, Z., Van Der Maaten, L., Weinberger, K.Q.: Densely Connected Convolutional Networks. In 2017 IEEE Conference on Computer Vision and Pattern Recognition (CVPR), pp. 2261–2269, Honolulu, HI, IEEE (2017)
14. Jiang, L., Huang, D., Liu, M., Yang, W.: Beyond Synthetic Noise: Deep Learning on Controlled Noisy Labels, August (2020). arXiv:1911.09781 [cs, stat]
15. Kim, B., Kim, H., Kim, K., Kim, S., Kim, J.: Learning Not to Learn: Training Deep Neural Networks With Biased Data. In 2019 IEEE/CVF Conference on Computer Vision and Pattern Recognition (CVPR), pp. 9004–9012, Long Beach, CA, USA, June IEEE (2019)
16. Kinyanjui, N.M., et al.: Estimating Skin Tone and Effects on Classification Performance in Dermatology Datasets. In Fair ML for Health, page 10, Vancouver, Canada, NeurIPS (2019)
17. Lio, P.A., Nghiem, P.: Interactive Atlas of Dermoscopy: 2000, Edra Medical Publishing and New Media. 208 pages. Journal of the American Academy of Dermatology. **50**(5), 807–808 (2004)
18. Paszke, A., et al.: PyTorch: An imperative style, high-performance deep learning library. In: Wallach, H., Larochelle, H., Beygelzimer, F. dAlché-Buc, E. Fox, and R. Garnett, Advances in Neural Information Processing Systems 32, pp. 8024–8035 Curran Associates Inc (2019)
19. Rotemberg, V., et al.: A patient-centric dataset of images and metadata for identifying melanomas using clinical context. Scientific Data. **8**(1) 34 (2021)
20. Szegedy, C., Vanhoucke, V., Ioffe, S., Shlens, J., Wojna, Z.: Rethinking the Inception Architecture for Computer Vision. In 2016 IEEE Conference on Computer Vision and Pattern Recognition (CVPR), pp. 2818–2826, IEEE Computer Society (2016)

21. Tan, M., Le, Q.V.: EfficientNet: Rethinking Model Scaling forConvolutional Neural Networks. Proceedings of the 36th International Conference on Machine Learning. **97** 6105–6114 (2019)
22. Tzeng, E., Hoffman, J., Darrell, T., Saenko, K.. Simultaneous Deep Transfer Across Domains and Tasks. In 2015 IEEE International Conference on Computer Vision (ICCV). pp. 4068–4076 (2015)
23. Xie, S., Girshick, R.B., Dollár, P., Tu, Z., He, K.: Aggregated Residual Transformations for Deep Neural Networks. In: 2017 IEEE Conference on Computer Vision and Pattern Recognition (CVPR) (2017)

Benchmarking and Boosting Transformers for Medical Image Classification

DongAo Ma[1], Mohammad Reza Hosseinzadeh Taher[1], Jiaxuan Pang[1], Nahid UI Islam[1], Fatemeh Haghighi[1], Michael B. Gotway[2], and Jianming Liang[1(✉)]

[1] Arizona State University, Tempe, AZ 85281, USA
{dongaoma,mhossei2,jpang12,nuislam,fhaghigh,jianming.liang}@asu.edu
[2] Mayo Clinic, Scottsdale, AZ 85259, USA
Gotway.Michael@mayo.edu

Abstract. Visual transformers have recently gained popularity in the computer vision community as they began to outrank convolutional neural networks (CNNs) in one representative visual benchmark after another. However, the competition between visual transformers and CNNs in medical imaging is rarely studied, leaving many important questions unanswered. As the first step, we benchmark how well existing transformer variants that use various (supervised and self-supervised) pre-training methods perform against CNNs on a variety of medical classification tasks. Furthermore, given the data-hungry nature of transformers and the annotation-deficiency challenge of medical imaging, we present a practical approach for bridging the domain gap between photographic and medical images by utilizing unlabeled large-scale in-domain data. Our extensive empirical evaluations reveal the following insights in medical imaging: (1) good initialization is more crucial for transformer-based models than for CNNs, (2) self-supervised learning based on masked image modeling captures more generalizable representations than supervised models, and (3) assembling a larger-scale domain-specific dataset can better bridge the domain gap between photographic and medical images via self-supervised continuous pre-training. We hope this benchmark study can direct future research on applying transformers to medical imaging analysis. All codes and pre-trained models are available on our GitHub page https://github.com/JLiangLab/BenchmarkTransformers.

Keywords: Vision Transformer · Transfer learning · Domain-adaptive pre-training · Benchmarking

1 Introduction

Visual transformers have recently demonstrated the potential to be considered as an alternative to CNNs in visual recognition. Though visual transformers

Supplementary Information The online version contains supplementary material available at https://doi.org/10.1007/978-3-031-16852-9_2.

K. Kamnitsas et al. (Eds.): DART 2022, LNCS 13542, pp. 12–22, 2022.
https://doi.org/10.1007/978-3-031-16852-9_2

have attained state-of-the-art (SOTA) performance across a variety of computer vision tasks [11,20], their architectures lack convolutional inductive bias, making them more data-hungry than CNNs [7,31]. Given the data-hungry nature of transformers and the challenge of annotation scarcity in medical imaging, the efficacy of existing visual transformers in medical imaging is unknown. Our preliminary analysis revealed that on medical target tasks with limited annotated data, transformers lag behind CNNs in random initialization (scratch) settings. To overcome the challenge of annotation dearth in medical imaging, transfer learning from ImageNet pre-trained models has become a common practice [9,10,18,35]. As such, the first question this paper seeks to answer is: *To what extent can ImageNet pre-training elevate transformers' performance to rival CNNs in medical imaging*?

Meanwhile, self-supervised learning (SSL) has drawn great attention in medical imaging due to its remarkable success in overcoming the challenge of annotation dearth in medical imaging [8,30]. The goal of the SSL paradigm is to learn general-purpose representations without using human-annotated labels [10,16]. Masked image modeling (MIM) methods, in addition to supervised pre-training, have recently emerged as promising SSL techniques for transformer models; the basic idea behind MIM-based methods is to learn representations by (randomly) masking portions of the input image and then recovering the input image at the masked areas. Recent advancements in MIM-based techniques have resulted in SSL techniques that outperform supervised pre-trained models in a variety of computer vision tasks [5,21]. As a result, the second question this paper seeks to answer is: *How generalizable are MIM-based self-supervised methods to medical imaging in comparison to supervised ImageNet pre-trained models*?

Furthermore, the *marked* differences between photographic and medical images [8,16,30] may result in a mismatch in learned features between the two domains, which is referred to as a "domain gap." Hosseinzadeh Taher *et al.* [16] recently demonstrated that using a CNN as the backbone, a *moderately-sized* medical image dataset is sufficient to bridge the domain gap between photographic and medical images via *supervised* continual pre-training. Motivated but different from this work and given the data-hungry nature of transformers, we investigate domain-adaptive pre-training in an SSL setting. Naturally, the third question this paper seeks to answer is: *How to scale up a domain-specific dataset for a transformer architecture to bridge the domain gap between photographic and medical images*?

In addressing the three questions, we conduct a benchmarking study to assess the efficacy of transformer-based models on numerous medical classification tasks involving different diseases (thorax diseases, lung pulmonary embolism, and tuberculosis) and modalities (X-ray and CT). In particular, (1) we investigate the importance of pre-training for transformers versus CNNs in medical imaging; (2) we assess the transferability of SOTA MIM-based self-supervised method to a diverse set of medical image classification tasks; and (3) we investigate domain-adaptive pre-training on large-scale photographic and medical images to tailor self-supervised ImageNet models for target tasks on chest X-rays.

Our extensive empirical study yields the following findings: (1) In medical imaging, good initialization is more vital for transformer-based models than for CNNs (see Fig. 1). (2) MIM-based self-supervised methods capture finer-grained representations that can be useful for medical tasks better than supervised pre-trained models (see Table 1). (3) Continuous self-supervised pre-training of the self-supervised ImageNet model on large-scale medical images bridges the domain gap between photographic and medical images, providing more generalizable pre-trained models for medical image classification tasks (see Table 2). We will contrast our study with related works in each subsection of Sect. 3 to show our novelties.

2 Benchmarking Setup

2.1 Transformer Backbones

In the target tasks in all experiments, we take two representative recent SOTA transformer backbones, including Vision Transformer (ViT) [7] and Swin Transformer (Swin) [22]. Visual transformer models, which have recently emerged as alternatives to convolutional neural networks (CNNs), have revolutionized computer vision fields. The groundbreaking work of ViT showcases how transformers can completely replace the CNNs backbone with a convolution-free model. Although ViT attains SOTA image classification performance, its architecture may not be suitable for use on dense vision tasks, such as object detection, segmentation, etc. Swin, a recent work, proposes a general-purpose transformer backbone to address this problem by building hierarchical feature maps, resulting in SOTA accuracy on object detection segmentation tasks. For transfer learning to the classification target tasks, we take the transformer pre-trained models and add a task-specific classification head. We assess the transfer learning performance of all pre-trained models by fine-tuning all layers in the downstream networks.

2.2 Target Tasks and Datasets

We consider a diverse suite of six common but challenging medical classification tasks including NIH ChestX-ray14 [32], CheXpert [17], VinDr-CXR [24], NIH Shenzhen CXR [19], RSNA PE Detection [6], and RSNA Pneumonia [1]. These tasks encompass various diseases (thorax diseases, lung pulmonary embolism, and tuberculosis) and modalities (X-ray and CT). We use official data split of these datasets if available; otherwise, we randomly divide the data into 80%/20% for training/testing. AUC (area under the ROC curve) is used to measure the performance of multi-label classification target tasks (NIH ChestX-ray14, CheXpert, and VinDr-CXR) and binary classification target tasks (NIH Shenzhen CXR and RSNA PE). Accuracy is used to evaluate multi-class classification target task (RSNA Pneumonia) performance. The mean and standard deviation of performance metrics over ten runs are reported in all experiments, and statistical analyses based on an independent two sample t-test are presented.

3 Benchmarking and Boosting Transformers

3.1 Pre-training is More Vital for Transformer-Based Models than for CNNs in Medical Imaging

Transformers have recently attained SOTA results and surpassed CNNs in a variety of computer vision tasks [11,20]. However, the lack of convolutional inductive bias in transformer architectures makes them more data-hungry than CNNs [7,31]. Therefore, to rival CNNs in vision tasks, transformers require a millions or even billions of labeled data [7,28,34]. Given the data-hungry nature of transformers and the challenge of annotation scarcity in medical imaging [10,25,27,35], it is natural to wonder whether transformers can compute with CNNs if they are used directly on medical imaging applications. Our preliminary analysis showed that in random initialization (scratch) settings, transformers lag behind CNNs on medical target tasks with limited annotated data. Taken together, we hypothesize that in medical imaging, transformers require pre-trained models to rival with CNNs. To put this hypothesis to the test, we empirically validate how well transformer variants (ViT-B and Swin-B) that use various (supervised and self-supervised) pre-training methods compete with CNNs on a range of medical classification tasks. In contrast to previous work [23] which only compared one transformer model with a CNN counterpart, we benchmark six newly-developed transformer models and three CNN models.

Experimental Setup. We evaluate the transferability of various popular transformer methods with officially released models on six diverse medical classification tasks. Our goal is to investigate the importance of pre-training for transformers versus CNNs in medical imaging. Given this goal, we use six popular transformer pre-trained models with ViT-B and Swin-B backbones and three standard CNNs pre-trained models with ResNet-50 backbones [15] that are already official and ready to use. Specifically, for supervised pre-training, we use official pre-trained ViT-B, Swin-B, and ResNet-50 on ImageNet-21K and pre-trained Swin-B and ResNet-50 on ImageNet-1K. For self-supervised pre-training, we use pre-trained ViT-B and Swin-B models with SimMIM [33] on ImageNet-1K, as well as pre-trained ViT-B and ResNet-50 models with MoCo v3 [4] on ImageNet-1K. The differences in pre-training data (ImageNet-1K or ImageNet-21K) are due to the availability of official pre-trained models.

Results and Analysis. Our evaluations in Fig. 1 suggest three major results. Firstly, in random initialization (scratch) settings (horizontal lines), transformers (*i.e.*, ViT-B and/or Swin-B) cannot compete with CNNs (*i.e.*, ResNet50) in medical applications, as they offer performance equally or even worse than CNNs. We attribute this inferior performance to transformers' lack of desirable inductive bias in comparison to CNNs, which has a negative impact on transformer performance on medical target tasks with limited annotated data. Secondly, Swin-B backbone consistently outperforms ViT-B across all target tasks.

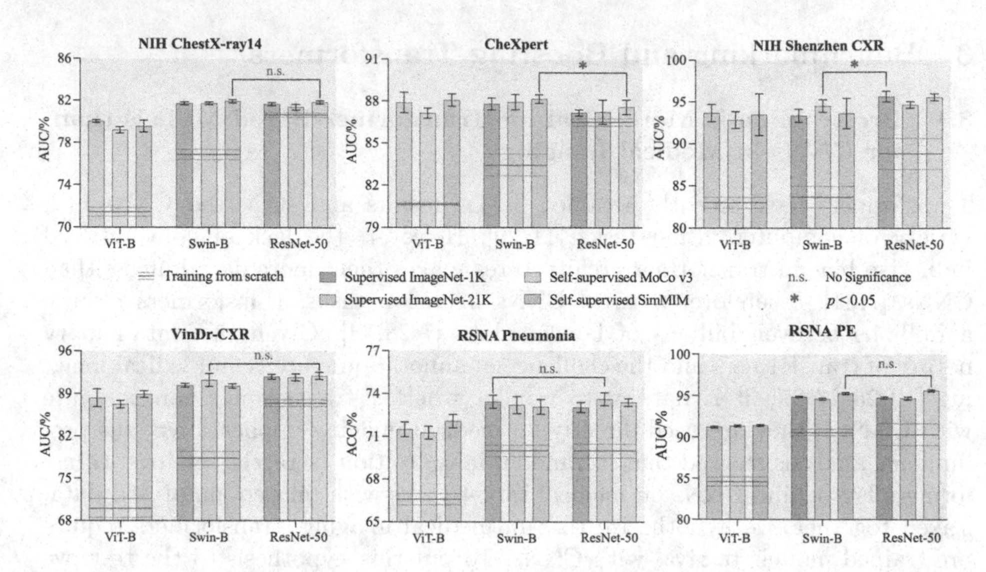

Fig. 1. In medical imaging, good initialization is more vital for transformer-based models than for CNNs. When training from scratch, transformers perform significantly worse than CNNs on all target tasks. However, with supervised or self-supervised pre-training on ImageNet, transformers can offer the same results as CNNs, highlighting the importance of pre-training when using transformers for medical imaging tasks. We conduct statistical analysis between the best of six pre-trained transformer models and the best of three pre-trained CNN models.

This reveals the importance of hierarchical inductive bias, which embedded in the Swin-B backbone, in elevating the performance of transformer-based models in medical image analysis. Thirdly, with supervised or self-supervised pre-training on ImageNet, transformers can offer competitive performance compare to CNNs, emphasizing the importance of pre-training when using transformers for medical imaging tasks. In particular, the best of six pre-trained transformer models outperform the best of three pre-trained CNN models in all target tasks, with the exception of NIH Shenzhen CXR, which can be attributed to a lack of sufficient training data (only 463 samples).

3.2 Self-supervised Learning Based on Masked Image Modeling is a Preferable Option to Supervised Baselines for Medical Imaging

Visual transformer models, while powerful, are prone to over-fitting and rely heavily on supervised pre-training on large-scale image datasets [7,34], such as JFT-300M [29] and ImageNet-21K [26]. In addition to supervised pre-training, self-supervised learning (SSL) techniques account for a substantial part of pre-trained transformer models. Masked Image Modeling (MIM) - an approach in which portions of the input image signals are randomly masked and then the original input signals are recovered at the masked area - has recently received great

Table 1. Self-supervised SimMIM model with the Swin-B backbone outperforms fully-supervised baselines. The best methods are bolded while the second best are underlined. For every target task, we conduct statistical analysis between the best (bolded) vs. others. Green-highlighted boxes indicate no statistically significant difference at the $p = 0.05$ level.

Initialization	Backbone	ChestX-ray14	CheXpert	Shenzhen	VinDr-CXR	RSNA Pneumonia	RSNA PE
Scratch	ViT-B	71.69±0.32	80.78±0.03	82.24±0.60	70.22±1.95	66.59±0.39	84.68±0.09
	Swin-B	77.04±0.34	83.39±0.84	92.52±4.98	78.49±1.00	70.02±0.42	90.63±0.10
Supervised	ViT-B	80.05±0.17	87.88±0.50	<u>93.67±1.03</u>	88.30±1.45	71.50±0.52	91.19±0.11
	Swin-B	<u>81.73±0.14</u>	87.80±0.42	93.35±0.77	**90.35±0.31**	<u>73.44±0.46</u>	<u>94.85±0.07</u>
SimMIM	ViT-B	79.55±0.56	<u>88.07±0.43</u>	93.47±2.48	88.91±0.55	72.08±0.47	91.39±0.10
	Swin-B	**81.95±0.15**	**88.16±0.31**	**94.12±0.96**	<u>90.24±0.35</u>	**73.66±0.34**	**95.27±0.12**

attention in computer vision for pre-training transformers in a self-supervised manner [14,33]. MIM-based self-supervised methods are widely accepted to capture more task-agnostic features than supervised pre-trained models, making them better suited for fine-tuning on various vision tasks [5,21]. We hypothesize that existing self-supervised transformer models pre-trained on photographic images will outperform supervised transformer models in the medical image domain, where there is a significant domain shift between medical and photographic images. To test this hypothesis, we consider two recent SOTA transformer backbones, ViT-B and Swin-B, and compare their supervised and self-supervised pre-trained models for various medical image classification tasks.

Experimental Setup. To investigate the efficacy of self-supervised and supervised pre-trained transformer models in medical imaging, we use existing supervised and SOTA self-supervised (*i.e.,* SimMIM) pre-trained models with two representative transformer backbones, ViT-B and Swin-B; all pre-trained models are fine-tuned on six different medical classification tasks. To provide a comprehensive evaluation, we also include results for the training of these two architectures from scratch. We use SimMIM instead of the concurrent MAE [14] as the representative MIM-based method because SimMIM has been demonstrated superior performance to MAE in medical image analysis [5].

Results and Analysis. As shown in Table 1, the self-supervised SimMIM model with the Swin-B backbone performs significantly better or on-par compared with both supervised baselines with either ViT-B or Swin-B backbones across all target tasks. The same observation of MIM-based models outperforming their supervised counterparts also exists in the finer-grained visual tasks, *i.e.,* object detection [21] and medical image segmentation [5]; different from them, we focus on coarse-grained classification tasks. Furthermore, we observe that the

SimMIM model with the Swin-B backbone consistently outperforms its counterpart with the ViT-B backbone in all cases, implying that the Swin-B backbone may be a superior option for medical imaging tasks to ViT-B. These findings suggest that the self-supervised SimMIM model with the Swin-B backbone could be a viable option for pre-training deep models in medical imaging applications.

3.3 Self-supervised Domain-Adaptive Pre-training on a Larger-Scale Domain-Specific Dataset Better Bridges the Domain Gap Between Photographic and Medical Imaging

Domain adaptation seeks to improve target model performance by reducing domain disparities between source and target domains. Recently, Hosseinzadeh Taher *et al.* [16] demonstrated that domain-adaptive pre-training can bridge the domain gap between natural and medical images. Particularly, Hosseinzadeh Taher *et al.* [16] first pre-trained a CNN model (*i.e.*, ResNet-50) on ImageNet and then on domain-specific datasets (*i.e.*, NIH ChestX-ray14 or CheXpert), demonstrating how domain-adaptive pre-training can tailor the ImageNet models to medical applications. Motivated by this work, we investigate domain-adaptive pre-training in the context of visual transformer architectures. Given the data-hungry nature of transformers and the annotation-dearth challenge of medical imaging, different from [16], we use the SSL pre-training approach to bridge the domain gap between photographic and medical images. Since no expert annotation is required in SSL pre-training, we are able to assemble multiple domain-specific datasets into a large-scale dataset, which is differentiated from Azizi *et al.* [2] who used only a single dataset.

Experimental Setup. We evaluate the transferability of five different self-supervised SimMIM models with the Swin-B backbone by utilizing three different pre-training datasets, including ImageNet, ChestX-ray14, and X-rays(926K)— a large-scale dataset that we created by collecting 926,028 images from 13 different chest X-ray datasets. To do so, we use SimMIM released ImageNet model as well as two models pre-trained on ChestX-ray14 and X-rays(926K) using SimMIM; additionally, we created two new models that were initialized through the self-supervised ImageNet pre-trained model followed by self-supervised pre-training on ChestX-ray14 (ImageNet→ChestX-ray14) and X-rays(926 K) (ImageNet→X-rays(926 K)). Every pre-training experiment trains for 100 epochs using the default SimMIM settings.

Results and Analysis. We draw the following observations from Table 2. (1) X-rays(926K) model consistently outperforms the ChestX-ray14 model in all cases. This observation suggests that scaling the pre-training data can significantly improve the self-supervised transformer models. (2) While the X-rays(926 K) model uses fewer images in the pre-training dataset than the ImageNet model, it shows superior or comparable performance over the ImageNet model across all

Table 2. The domain-adapted pre-trained model which utilized a large number of in-domain data (X-rays(926K)) in an SSL manner achieves the best performance across all five target tasks. The best methods are bolded while the second best are underlined. For each target task, we conducted the independent two sample t-test between the best (bolded) vs. others. The absence of a statistically significant difference at the $p = 0.05$ level is indicated by green-highlighted boxes.

Initialization	ChestX-ray14	CheXpert	Shenzhen	VinDr-CXR	RSNA Pneumonia
Scratch	77.04±0.34	83.39±0.84	83.92±1.19	78.49±1.00	70.02±0.42
ImageNet	81.95±0.15	88.16±0.31	93.63±1.80	90.24±0.35	73.66±0.34
ChestX-ray14	78.87±0.69	86.75±0.96	93.03±0.48	79.86±1.82	71.99±0.55
X-rays(926K)	82.72±0.17	87.83±0.23	95.21±1.44	90.60±1.95	73.73±0.50
ImageNet→ChestX-ray14	82.45±0.15	87.74±0.31	94.83±0.90	90.33±0.88	73.87±0.48
ImageNet→X-rays(926K)	**83.04±0.15**	**88.37±0.40**	**95.76±1.79**	**91.71±1.04**	**74.09±0.39**

target tasks. In line with Hosseinzadeh Taher *et al.* [16], this implies that, whenever possible, in-domain medical transfer learning should be preferred over ImageNet transfer learning. (3) The overall trend highlights the benefit of domain-adaptive pre-training, which leverages the ImageNet model's learning experience and further refines it with domain-relevant data. Specifically, fine-tuning both domain-adapted models (ImageNet→ChestX-ray14 and ImageNet→X-rays(926 K)) outperforms ImageNet and corresponding in-domain models in all target tasks, with one exception; in the CheXpert, the ImageNet→ChestX-ray14 model performs worse in CheXpert than the corresponding ImageNet model. This exception, in line with Hosseinzadeh Taher *et al.* [16], suggests that the in-domain pre-training dataset should be larger than the target dataset. It is noteworthy that this gap was filled later by ImageNet→X-rays(926 K) model, which utilized more in-domain data. This highlights the significance of larger-scale medical data in improving the transformers' ability to learn more discriminative representations.

4 Conclusion and Future Work

We manifest an up-to-date benchmark study to shed light on the efficacy and limitations of existing visual transformer models in medical image classification when compared to CNN counterparts. Our extensive experiments yield important findings: (1) a good pre-train model can allow visual transformers to compete with CNNs on medical target tasks with limited annotated data; (2) MIM-based self-supervised methods, such as SimMIM, play an important role in pre-training visual transformer models, preventing them from over-fitting in medical target tasks; and (3) assembling multiple domain-specific datasets into a larger-scale one can better bridge the domain gap between photographic and medical imaging via continual SSL pre-training.

Future Work: Recently, many transformer-based UNet architectures have been developed for 3D medical image segmentation [3,5,12,13]. To make a comprehensive benchmarking study of transformers for medical image analysis, we will extend the evaluation to more modalities and medical image segmentation tasks in future work.

Acknowledgments. This research has been supported in part by ASU and Mayo Clinic through a Seed Grant and an Innovation Grant, and in part by the NIH under Award Number R01HL128785. The content is solely the responsibility of the authors and does not necessarily represent the official views of the NIH. This work has utilized the GPUs provided in part by the ASU Research Computing and in part by the Extreme Science and Engineering Discovery Environment (XSEDE) funded by the National Science Foundation (NSF) under grant numbers: ACI-1548562, ACI-1928147, and ACI-2005632. We thank Manas Chetan Valia and Haozhe Luo for evaluating the pre-trained ResNet50 models on the five chest X-ray tasks and the pre-trained transformer models on the VinDr-CXR target tasks, respectively. The content of this paper is covered by patents pending.

References

1. Rsna pneumonia detection challenge (2018). https://www.kaggle.com/c/rsna-pneumonia-detection-challenge
2. Azizi, S., et al.: Big self-supervised models advance medical image classification. In: Proceedings of the IEEE/CVF International Conference on Computer Vision, pp. 3478–3488 (2021)
3. Cao, H., et al.: Swin-unet: Unet-like pure transformer for medical image segmentation. arXiv preprint arXiv:2105.05537 (2021)
4. Chen, X., Xie, S., He, K.: An empirical study of training self-supervised vision transformers. In: Proceedings of the IEEE/CVF International Conference on Computer Vision, pp. 9640–9649 (2021)
5. Chen, Z., et al.: Masked image modeling advances 3D medical image analysis. arXiv preprint arXiv:2204.11716 (2022)
6. Colak, E., et al.: The RSNA pulmonary embolism CT dataset. Radiol. Artif. Intell. **3**(2) (2021)
7. Dosovitskiy, A., et al.: An image is worth 16x16 words: Transformers for image recognition at scale. arXiv preprint arXiv:2010.11929 (2020)
8. Haghighi, F., Hosseinzadeh Taher, M.R., Gotway, M.B., Liang, J.: DiRA: Discriminative, restorative, and adversarial learning for self-supervised medical image analysis. In: Proceedings of the IEEE/CVF Conference on Computer Vision and Pattern Recognition (CVPR), pp. 20824–20834 (2022)
9. Haghighi, F., Hosseinzadeh Taher, M.R., Zhou, Z., Gotway, M.B., Liang, J.: Learning semantics-enriched representation via self-discovery, self-classification, and self-restoration. In: Martel, A.L. (ed.) MICCAI 2020. LNCS, vol. 12261, pp. 137–147. Springer, Cham (2020). https://doi.org/10.1007/978-3-030-59710-8_14
10. Haghighi, F., Taher, M.R.H., Zhou, Z., Gotway, M.B., Liang, J.: Transferable visual words: exploiting the semantics of anatomical patterns for self-supervised learning. IEEE Trans. Med. Imaging **40**(10), 2857–2868 (2021). https://doi.org/10.1109/TMI.2021.3060634

11. Han, K., et al.: A survey on vision transformer. IEEE Trans. Patt. Anal. Mach. Intell. (2022)
12. Hatamizadeh, A., Nath, V., Tang, Y., Yang, D., Roth, H., Xu, D.: Swin UNETR: Swin transformers for semantic segmentation of brain tumors in MRI images. arXiv preprint arXiv:2201.01266 (2022)
13. Hatamizadeh, A., et al.: UNETR: Transformers for 3D medical image segmentation. In: Proceedings of the IEEE/CVF Winter Conference on Applications of Computer Vision, pp. 574–584 (2022)
14. He, K., Chen, X., Xie, S., Li, Y., Dollár, P., Girshick, R.: Masked autoencoders are scalable vision learners. In: Proceedings of the IEEE/CVF Conference on Computer Vision and Pattern Recognition, pp. 16000–16009 (2022)
15. He, K., Zhang, X., Ren, S., Sun, J.: Deep residual learning for image recognition. In: Proceedings of the IEEE Conference on Computer Vision and Pattern Recognition, pp. 770–778 (2016)
16. Hosseinzadeh Taher, M.R., Haghighi, F., Feng, R., Gotway, M.B., Liang, J.: A systematic benchmarking analysis of transfer learning for medical image analysis. In: Albarqouni, S. (ed.) DART/FAIR -2021. LNCS, vol. 12968, pp. 3–13. Springer, Cham (2021). https://doi.org/10.1007/978-3-030-87722-4_1
17. Irvin, J., et al.: Chexpert: A large chest radiograph dataset with uncertainty labels and expert comparison. In: Proceedings of the AAAI Conference on Artificial Intelligence, vol. 33, pp. 590–597 (2019)
18. Islam, N.U., Gehlot, S., Zhou, Z., Gotway, M.B., Liang, J.: Seeking an optimal approach for computer-aided pulmonary embolism detection. In: Lian, C., Cao, X., Rekik, I., Xu, X., Yan, P. (eds.) MLMI 2021. LNCS, vol. 12966, pp. 692–702. Springer, Cham (2021). https://doi.org/10.1007/978-3-030-87589-3_71
19. Jaeger, S., Candemir, S., Antani, S., Wáng, Y.X.J., Lu, P.X., Thoma, G.: Two public chest x-ray datasets for computer-aided screening of pulmonary diseases. Quant. Imaging Med. Surg. **4**(6), 475 (2014)
20. Khan, S., Naseer, M., Hayat, M., Zamir, S.W., Khan, F.S., Shah, M.: Transformers in vision: A survey. ACM Computing Surveys (CSUR) (2021)
21. Li, Y., Xie, S., Chen, X., Dollar, P., He, K., Girshick, R.: Benchmarking detection transfer learning with vision transformers. arXiv preprint arXiv:2111.11429 (2021)
22. Liu, Z., et al.: Swin transformer: Hierarchical vision transformer using shifted windows. In: Proceedings of the IEEE/CVF International Conference on Computer Vision, pp. 10012–10022 (2021)
23. Matsoukas, C., Haslum, J.F., Söderberg, M., Smith, K.: Is it time to replace CNNs with transformers for medical images? arXiv preprint arXiv:2108.09038 (2021)
24. Nguyen, H.Q., et al.: VinDR-CXR: An open dataset of chest x-rays with radiologist's annotations. arXiv preprint arXiv:2012.15029 (2020)
25. Parvaiz, A., Khalid, M.A., Zafar, R., Ameer, H., Ali, M., Fraz, M.M.: Vision transformers in medical computer vision-a contemplative retrospection. arXiv preprint arXiv:2203.15269 (2022)
26. Russakovsky, O., et al.: ImageNet large scale visual recognition challenge. Int. J. Comput. Vision **115**(3), 211–252 (2015)
27. Shamshad, F., et al.: Transformers in medical imaging: A survey. arXiv preprint arXiv:2201.09873 (2022)
28. Steiner, A., Kolesnikov, A., Zhai, X., Wightman, R., Uszkoreit, J., Beyer, L.: How to train your viT? data, augmentation, and regularization in vision transformers. arXiv preprint arXiv:2106.10270 (2021)

29. Sun, C., Shrivastava, A., Singh, S., Gupta, A.: Revisiting unreasonable effectiveness of data in deep learning era. In: Proceedings of the IEEE International Conference on Computer Vision, pp. 843–852 (2017)
30. Taher, M.R.H., Haghighi, F., Gotway, M.B., Liang, J.: CAid: Context-aware instance discrimination for self-supervised learning in medical imaging. arXiv:2204.07344 (2022). https://doi.org/10.48550/ARXIV.2204.07344, https://arxiv.org/abs/2204.07344
31. Touvron, H., Cord, M., Douze, M., Massa, F., Sablayrolles, A., Jégou, H.: Training data-efficient image transformers & distillation through attention. In: International Conference on Machine Learning, pp. 10347–10357, PMLR (2021)
32. Wang, X., Peng, Y., Lu, L., Lu, Z., Bagheri, M., Summers, R.M.: Chestx-ray8: Hospital-scale chest x-ray database and benchmarks on weakly-supervised classification and localization of common thorax diseases. In: Proceedings of the IEEE Conference on Computer Vision and Pattern Recognition. pp. 2097–2106 (2017)
33. Xie, Z., et al.: SimMIM: A simple framework for masked image modeling. arXiv preprint arXiv:2111.09886 (2021)
34. Zhai, X., Kolesnikov, A., Houlsby, N., Beyer, L.: Scaling Vision Transformers. arXiv preprint arXiv:2106.04560 (2021)
35. Zhou, Z., Sodha, V., Pang, J., Gotway, M.B., Liang, J.: Models genesis. Med. Image Anal. **67**, 101840 (2021)

Supervised Domain Adaptation Using Gradients Transfer for Improved Medical Image Analysis

Shaya Goodman[1], Hayit Greenspan[1], and Jacob Goldberger[2(✉)]

[1] Tel-Aviv University, Tel-Aviv, Israel
yishayahug@mail.tau.ac.il, hayit@eng.tau.ac.il
[2] Bar-Ilan University, Ramat-Gan, Israel
jacob.goldberger@biu.ac.il

Abstract. A well known problem in medical imaging is the performance degradation that occurs when using a model learned on source data, in a new site. Supervised Domain Adaptation (SDA) strategies that focus on this challenge, assume the availability of a limited number of annotated samples from the new site. A typical SDA approach is to pre-train the model on the source site and then fine-tune on the target site. Current research has thus mainly focused on which layers should be fine-tuned. Our approach is based on transferring also the *gradients history* of the pre-training phase to the fine-tuning phase. We present two schemes to transfer the gradients information to improve the generalization achieved during pre-training while fine-tuning the model. We show that our methods outperform the *state-of-the-art* with different levels of data scarcity from the target site, on multiple datasets and tasks.

Keywords: Transfer learning · Site adaptation · Gradient transfer · MRI segmentation · Xray classification

1 Introduction

Deep learning has achieved remarkable success in many computer vision and medical imaging tasks, but current methods often rely on a large amount of labeled training data [5,9]. *Transfer learning* is commonly used to compensate for the lack of sufficient training data in a given target task by using knowledge from a related source task. Whenever a model, trained on data from one distribution, is given data belonging to a slightly different distribution, a detrimental effect can occur that decreases the inference quality [20]. This problem is especially acute in the field of medical imaging since any data collection instance (e.g., an MRI apparatus) might belong to a domain of its own due to the peculiarities of a model or the scanning protocol [3].

In the current study we tackle the following setup: we have a sufficient amount of labeled MRI data taken from a source site but only a small number of patients

This research was supported by the Ministry of Science & Technology, Israel.

K. Kamnitsas et al. (Eds.): DART 2022, LNCS 13542, pp. 23–32, 2022.
https://doi.org/10.1007/978-3-031-16852-9_3

with labeled MRI data from the target site. This setup is called *supervised domain adaptation*. Another setup is unsupervised domain adaptation where we only have unlabeled data from the target site (see e.g. [4,7]).

Fine-tuning is the most widely used approach to transfer learning when working with deep learning models. Given a pre-trained model which was trained on the source site, it is further trained on the data from the target site.

Recent research in the area of domain shift mainly deals with which layers should be trained in the fine-tuning phase. The earlier approaches were based on the work of Yosinski et al. [22], which suggests fine-tuning the layers at the end of the network. Later approaches use the procedure in [1], which argues that medical images (e.g. MRI) mostly contain low-level domain shift, so that the first layers should be targeted. Shirokikh et al. [16] compared fine-tuning of the first layers to fine-tuning of the last layers and the whole network. The findings show that fine-tuning of the first layers is superior to fine-tuning of the last ones and is preferable to fine-tuning of the whole network in the case of annotated target data scarcity. Zakazov et al. [23] recently presented a domain adaptation method which learns a policy to choose between fine-tuning and reusing pre-trained network layers. Few methods such as [11,15] use the source model as an $L2$ regularization that minimizes the distance from the pretrained model in the fine-tuning phase. In this work we suggest a new fine-tuning method. In addition to transferring the model parameters from the pre-training phase, we transfer the *model gradients* that were obtained at the end of the learning process. By doing that, we transfer the information on the network state in the parameter space. We show that our approach outperforms the state-of-the-art SDA methods both in segmentation and classification medical imaging tasks.[1]

2 Transferring the Parameter Gradients

Most supervised domain adaptation methods are based on transferring the model weights from the source to the target domain and then fine-tuning them on a small amount of labeled data from the target domain. We propose passing more information from the source training phase to the target training phase rather than only the model parameters. Let $\theta \in \Omega$ be a network parameter-set trained on labeled data from the source task $(x_1, y_1), ..., (x_n, y_n)$. The gradient of the loss function with respect to the weights of the network is:

$$L'(\theta) = \frac{1}{n} \sum_{t=1}^{n} \frac{d}{d\theta} l(x_t, y_t, \theta), \qquad \theta \in \Omega \tag{1}$$

where l is the single instance loss function (e.g. cross entropy). L is minimized during the training process but its gradient is not necessarily set to zero at the end of training due to the non-convexity of the loss function, regularization constraints, validation-based early stopping, a limit on the number of training

[1] The code to reproduce our experiments is available at: https://github.com/yishayahu/MMTL.git.

epochs, etc. The parameter-set θ and its gradient $L'(\theta)$ define the state of the network in the parameter space in a way similar to the position and velocity that completely define the state of a system in classical mechanics.

We examine the impact of transferring $L'(\theta)$ along with θ from the pre-training phase to the fine-tuning phase. In practice, the gradient $L'(\theta)$ is never explicitly computed during training. Instead, an approximated gradient, computed on mini-batches, is used to update the weights. Updating the weights by the approximated gradient can be a noisy process, since the approximated gradient can point to the opposite direction from the exact gradient. To limit the effect of this process, an exponentially decayed moving average of gradients obtained from previous mini-batches is maintained. This helps in gradually forgetting old weights (when the network was poorly trained) and averaging the recent weights (to get rid of the noise).

We propose two variants of gradient transfer that are both based on the mini-batch computation of the gradient: The first gradient transfer method is the *Mixed Minibatch Transfer Learning* (MMTL) method. It is based on gradually exposing the optimization process to a larger portion of the target samples, while continuing to compute the gradient on samples from the source domain. The gradient that is used in the loss minimization process is obtained as:

$$\alpha L'_{source}(\theta) + (1-\alpha)L'_{target}(\theta) = \frac{1}{n}\left(\sum_{t=1}^{n_1} \frac{d}{d\theta} l(x_t, y_t, \theta) + \sum_{t=1}^{n-n_1} \frac{d}{d\theta} l(x'_t, y'_t, \theta)\right)$$

where a fraction $\alpha = n_1/n$ of the n mini-batch data-points (x_t, y_t) is taken from the source domain and the rest (x'_t, y'_t) are taken from the target domain. This smooth transition from the source domain to the target domain differs from the standard fine-tuning procedure in which there is a sharp transition from training on the source domain to training on the target domain. In the suggested process, more information is transferred from the pre-training phase to the fine-tuning phase. MMTL is summarized in Algorithm box 1.

Algorithm 1. Mixed Minibatch based Transfer Learning (MMTL)

$\theta_0 \leftarrow \theta_{pretrained}$
for t in 1 to num_of_steps **do**
  $n\_target \leftarrow int(\frac{current\_epoch}{num\_epochs} \times batch\_size)$
  $n\_source \leftarrow batch\_size - n\_target$
  $data \leftarrow concat(sample(source, n\_source), sample(target, n\_target))$
  $g_t \leftarrow \nabla f_t(\theta_{t-1}, data)$
  $\theta_t \leftarrow optimizer.step(\theta_{t-1}, g_t)$
end for

The second gradient transfer method takes the gradients history that was acquired during the pre-training phase and injects this information into the fine-tuning optimizer. The implementation specifics can differ as a function of the optimizer used. We applied this concept to two standard optimizers: Stochastic Gradient Descent (SGD) and Adaptive Moment Estimation (Adam) [8]. In order to keep the gradient history for a significant part of the fine-tuning phase, we need to adjust the exponential-decay parameters. In our experiments, we used $\mu = 0.99$ in SGD and $\beta_1, \beta_2 = 0.99, 0.999$ in Adam. Furthermore, the Adam optimizer uses a heuristic that gives more weight to the first training steps in the history calculation to avoid the tendency to zero. If we start the training from the first step it will quickly forget the pre-trained history. On the other hand, if we start from the last pre-training step (6K in our case) this might give the pre-trained history too much impact and undermine the fine-tuning procedure. Thus, in our experiments, we consider the starting point of the target training as step 100. We dub this method *Optimizer-Continuation Transfer Learning* (OCTL). The implementation details of the application of OCTL to SGD and Adam are shown in Table 1.

Table 1. Optimizer-Continuation Transfer Learning (OCTL) algorithm

Algorithm 2. OCTL - SGD

$\theta_0 \leftarrow \theta_{pretrained}$
$H_0 \leftarrow H_{pretrained}$ (SGD gradients history)
$\mu \leftarrow 0.99$
$\gamma \leftarrow learning\_rate$
for t in 1 to num_of_steps **do**
  $g_t \leftarrow \nabla f_t(\theta_{t-1})$
  $H_t \leftarrow \mu H_{t-1} + g_t$
  $\theta_t \leftarrow \theta_{t-1} - \gamma H_t$
end for

Algorithm 3. OCTL - Adam

$\theta_0 \leftarrow \theta_{pretrained}$
$M_0 \leftarrow M_{pretrained}$ (gradients first moment history)
$V_0 \leftarrow V_{pretrained}$ (gradients second moment history)
$\beta_1, \beta_2 \leftarrow 0.99, 0.999$
$\gamma \leftarrow learning\_rate$
for t in 1 to num_of_steps **do**
  $g_t \leftarrow \nabla f_t(\theta_{t-1})$
  $M_t \leftarrow \beta_1 M_{t-1} + (1 - \beta_1) g_t$
  $V_t \leftarrow \beta_2 V_{t-1} + (1 - \beta_2) g_t^2$
  $\bar{M}_t \leftarrow M_t / (1 - \beta_1^{(t+100)})$
  $\bar{V}_t \leftarrow V_t / (1 - \beta_2^{(t+100)})$
  $\theta_t \leftarrow \theta_{t-1} - \gamma \bar{M}_t / (\sqrt{\bar{V}_t} + \epsilon)$
end for

In the next section we evaluate the performance of the two gradient transfer procedures described above. We used the same hyper-parameter configuration described above in all our experiments. This demonstrates the robustness of our method across different tasks and datasets.

3 Experiments and Results

To examine the performance and robustness of the suggested domain adaptation methods we evaluated them on three different data-sets, two segmentation tasks and one classification task. All our experiments were conducted considering the following scenario: we have a sufficient amount of labeled data taken from a source site and small amount of labeled data from the target site.

We evaluate the following site adaptation methods:

- *Fine-tune-all* (FTA): fine-tune the model parameters using several training epochs on the target site data [16].
- *Fine-tune-first* (FTF): fine-tune the first layers of the pre-trained model [16].
- *L2SP*: fine-tune the model using a regularization term for minimizing the L2-distance to the pretrained model parameters [11].
- *Spottunet* (ST): fine-tune the model using the method described in [23] (which is the *state-of-the-art* for the CC359 dataset [17]).
- Our gradient transfer methods: MMTL and OCTL.

All compared methods start with the same network that was trained on the source site data. As an upper bound for the performance, we also directly trained a network on the target site using all the available training data.

When comparing to *unsupervised* domain adaptation methods (such as AdaptSegNet [19]), all the *supervised* methods mentioned above yield significantly better results (as expected), thus we exclude comparison to unsupervised methods.

3.1 MRI Skull Stripping

Data. The publicly available dataset CC359 [17] consists of 359 MR images of heads and the task is skull stripping. The dataset was collected from six sites which exhibit domain shift resulting in a severe score deterioration [16]. The data preprocessing steps were: interpolation to $1 \times 1 \times 1$ mm voxel spacing, and scaling intensities into 0 to 1 interval. Each patient yielded approx. 270 slices. We compared the adaptation methods using 90, 270, 540 and 1080 slices out of the total approx. 12K slices of the target site (by including 1, 1, 2 and 4 patients out of 45). To evaluate domain adaptation approaches, we used the surface Dice score [14] at a tolerance of 1 mm. While preserving the consistency with the methodology of [16], we found surface Dice score to be a more suitable metric for the brain segmentation task than the standard Dice Score (similar to [23]).

Architecture and Training. The experimental evaluation provided in [16] shows that neither architecture nor training procedure variations, e.g. augmentation, affect the relative performance of conceptually different approaches. Therefore, in all our experiments we used the 2D U-Net architecture implementation from [23]. In all the experiments we minimized the binary Cross-Entropy loss. All the methods were trained with a batch size 16 for 60 epochs with the learning rate of 10^{-3} reduced to 10^{-4} at the 45-th epoch. We ensured that all the models reached the loss plateau.

Results. The segmentation results using Adam and SGD optimizers are shown in Table 2. We used three different random seeds. For each seed, we tested the methods over all the 30 possible pairs of single-source and single-target sites and calculated the mean. Finally, we reported the mean and the std of the three mean values. Table 2 shows that MMTL maximized the chosen metric. We observed that the OCTL method was only effective in cases for which we had a relatively large amount of data from the target task. This method may be too aggressive, and it keeps the model too close to the source minimum. The average results in case we use all training data (12000 slices) to directly train the target site were 94.22 (Adam) and 84.61 (SGD). The average results without any fine-tuning were 51.94 (Adam) and 54.63 (SGD).

Table 2. Segmentation sDice results on the MRI skull stripping task using the SGD optimizer (top) and Adam (bottom).

Num. slices	FTA [16]	FTF [16]	L2SP [11]	ST [23]	Gradient transfer	
					OCTL	MMTL
90	74.45 ± 1.08	73.23 ± 0.57	73.69 ± 0.27	75.29 ± 1.11	73.44 ± 0.41	**78.46 ± 0.72**
270	76.41 ± 0.46	73.42 ± 0.88	74.83 ± 1.11	75.72 ± 0.65	75.66 ± 0.43	**79.63 ± 0.56**
540	77.31 ± 0.09	74.21 ± 0.92	74.68 ± 0.79	76.92 ± 0.31	80.50 ± 1.68	**81.34 ± 1.06**
1080	82.64 ± 0.56	77.49 ± 0.79	79.10 ± 1.08	81.32 ± 1.32	**86.98 ± 0.83**	84.16 ± 0.30
90	80.45 ± 0.46	84.36 ± 0.27	88.41 ± 0.13	86.09 ± 1.24	81.25 ± 0.23	**89.13 ± 0.45**
270	81.96 ± 1.47	86.27 ± 0.47	88.34 ± 0.63	86.31 ± 0.92	81.98 ± 0.58	**89.36 ± 0.82**
540	85.98 ± 0.33	87.67 ± 0.50	88.62 ± 0.43	87.95 ± 0.81	88.44 ± 0.86	**90.76 ± 0.77**
1080	89.37 ± 0.19	90.17 ± 1.33	89.20 ± 0.76	89.10 ± 0.85	90.76 ± 0.39	**91.65 ± 0.38**

The MMTL scheme makes the transition from the source site to the target site smoother. It prevents the model from overfitting the few labeled training examples from the target site. Instead, it preserves the information acquired throughout the pre-training phase. Figure 1 visualizes the feature representation of the model for every patient in CC359 [17] during the training process. To visualize the model's feature representation we applied dimensional reduction using t-SNE on the bottleneck layer of the U-Net. During the FTA model training process (left panel) the train slices and most of the test slices were separated into two different clusters, while the cluster that contains the training examples has much lower loss. This over-fitting behavior was not present in MMTL (right panel) since it learned from the source training set.

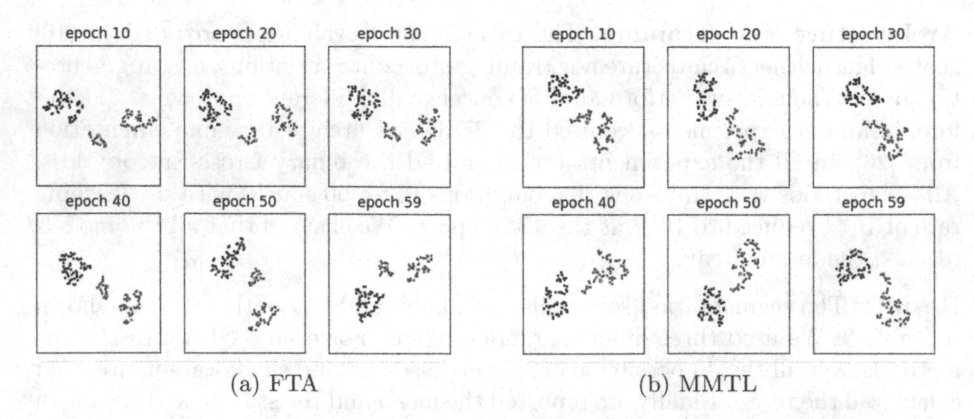

Fig. 1. Progress of the transfer learning training: source site examples (Blue), target site train examples (Green), target site test examples (Red). (Color figure online)

Table 3. Segmentation sDice mean results on the MRI skull stripping date-set computed on the source site using the model that was fine-tuned on target site. The results are averaged over all 30 pairs of source-target sites.

Number of slices	Source Model	Fine tuned model				
		FTF [16]	L2SP [11]	ST [23]	OCTL	MMTL
90	84.61	58.42	59.91	57.64	52.07	**86.99**
270		57.48	62.76	58.70	54.59	**86.74**
540		53.06	55.43	53.12	54.10	**85.83**
1080		55.78	59.25	55.45	60.09	**85.59**

Site adaptation focuses on the results of the fine-tuned model on the target site. Another question is how well this model addresses "catastrophic forgetting", i.e., how much does the performance drop on source data after fine tuned on target data. Table 3 shows the mean sDice score of the target model when applied on the source site. We can see that MMTL manages to maintain the performance of the source model and even it is slightly better. In contrast, in all other compared methods there is a strong performance degradation.

3.2 Prostate MRI Segmentation

Data. We use a publicly available multi-site dataset for prostate MRI segmentation which contains prostate T2-weighted MRI data (with segmentation masks) collected from 6 different data sources with a distribution shift out of three public datasets. Details of the data and imaging protocols from the six different sites appear in [13]. Samples of sites A and B are from the NCI-ISBI13 dataset [2], samples of site C are from the I2CVB dataset [10], and samples of sites D, E and F are from PROMISE12 dataset [12]. For pre-processing, we normalized each sample to have zero mean and unit variance in intensity value before inputting to the network. In this scenario we have a multi-source single-target setup: we used five sites as the source and the last site as the target. The results are calculated on six possible target sites using three different random seeds. To evaluate the different approaches, we used the Dice Score. We employed the same network architecture and training procedure as in the experiment described above. In addition we used the same hyper-parameters that were found in the previous experiment. The results are shown in Table 4. We can see that MMTL outperforms the other methods. The average results without any fine-tuning were 69.80. Qualitative segmentation results are shown at Fig. 2.

Table 4. Segmentation Dice on prostate MRI dataset using the Adam optimizer.

Num. slices	FTA [16]	FTF [16]	L2SP [11]	ST [23]	Gradient transfer	
					OCTL	MMTL
10	81.36 ± 0.64	77.98 ± 0.38	80.47 ± 0.12	80.01 ± 0.20	81.24 ± 0.44	$\mathbf{83.15 \pm 0.41}$
20	82.38 ± 0.57	77.72 ± 0.63	82.99 ± 0.83	81.79 ± 0.56	82.27 ± 0.49	$\mathbf{83.98 \pm 0.33}$
40	84.27 ± 0.83	83.35 ± 0.65	83.58 ± 0.70	84.07 ± 0.11	84.09 ± 0.72	$\mathbf{85.70 \pm 0.91}$

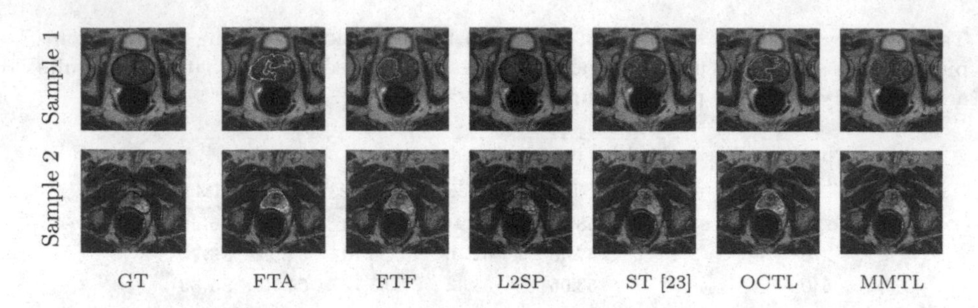

Fig. 2. Two qualitative segmentation results from the prostate MRI dataset.

Table 5. Mean AUC results on transfer learning from cheXpert [6] to ChestX-ray14 [21] (top) and from ChestX-ray14 to cheXpert (bottom).

Num. slices	FTA [16]	FTF [16]	L2SP [11]	ST [23]	Gradient transfer	
					OCTL	MMTL
128	64.13 ± 0.28	65.37 ± 0.09	67.93 ± 0.17	64.61 ± 0.33	65.87 ± 0.09	**68.12 ± 0.24**
256	66.79 ± 0.20	67.65 ± 0.57	68.20 ± 0.25	66.04 ± 0.27	67.86 ± 0.12	**69.85 ± 0.04**
512	67.45 ± 0.93	68.07 ± 0.08	68.38 ± 0.23	67.30 ± 0.09	68.86 ± 0.08	**70.85 ± 0.04**
1024	67.27 ± 0.60	70.23 ± 0.50	68.83 ± 0.17	68.42 ± 0.30	71.20 ± 0.32	**73.57 ± 0.71**
128	62.12 ± 0.04	63.13 ± 0.19	**65.82 ± 0.04**	61.86 ± 0.41	63.95 ± 0.21	64.74 ± 0.08
256	64.81 ± 0.24	66.91 ± 0.25	67.44 ± 0.22	64.73 ± 0.07	67.01 ± 0.19	**68.16 ± 0.28**
512	65.93 ± 0.46	69.01 ± 0.04	67.78 ± 0.17	67.02 ± 0.56	68.91 ± 0.56	**69.45 ± 0.07**
1024	68.55 ± 0.35	70.21 ± 0.15	68.17 ± 0.08	69.93± 0.14	70.96 ± 0.19	**71.28 ± 0.08**

3.3 Chest X-ray Multi-label Datasets

We apply our method to a transfer learning problem from one X-ray multi-label classification dataset to another. In this case, the label-sets in the two datasets are not the same (unlike site adaptation) and, hence, there is a different classification head for each task that we need to train from scratch.

Data. CheXpert [6] is a large public dataset for chest images, consisting of 224K chest radiographs from 65K patients. Each image was labeled for the presence of 14 observations as positive, negative, or uncertain. ChestX-ray14 [21] is another dataset that contains 112K X-ray images from 30K patients, labeled with up to 14 different thoracic diseases.

Architecture and training. We used Densenet121 architecture trained with the Adam optimizer with a learning rate of 1e-4 both for pre-training and fine-tuning. We randomly split the data to train, validation and test. We applied the compared methods to fine-tune on 128, 256, 512 and 1024 images, where 1024 is less than 0.1% of the training data.

Results. To evaluate the performance we used the Mean Area Under the ROC Curve measure. Table 5 shows mean and standard deviation of three different random seeds. It can be seen that our approach outperforms previous methods.

We finally note that a recent paper [18] proposed a similar idea. They, however, addressed the problem of transfer learning between different tasks where there is a need for a parameter tuning on a validation set from the target task. In our case we address the same task in different domains and no parameter tuning is needed. In addition we also improve the performance on the source domain.

To conclude, we propose a general domain adaptation concept of transferring the pre-trained gradients to the fine-tuning phase along with the parameters. This concept was implemented by two methods - OCTL and MMTL, We found that MMTL consistently yields better results than OCTL and also consistently outperforms all other compared methods. The presented experiments demonstrate that our methods manage to transfer the information from the pre-trained model while fine-tuning the model, in both segmentation and classification medical imaging tasks.

References

1. Aljundi, R., Tuytelaars, T.: Lightweight unsupervised domain adaptation by convolutional filter reconstruction (2016)
2. Bloch, N., et al.: NCI-ISBI 2013 challenge: automated segmentation of prostate structures. The Cancer Imaging Archive 370 (2015)
3. Glocker, B., Robinson, R., de Castro, D.C., Dou, Q., Konukoglu, E.: Machine learning with multi-site imaging data: An empirical study on the impact of scanner effects. CoRR abs/1910.04597 (2019)
4. Goodman, S., Kasten-Serlin, S., Greenspan, H., Goldberger, J.: Unsupervised site adaptation by intra-site variability alignment. In: MICCAI Workshop on Domain Adaptation and Representation Transfer (DART) (2022)
5. He, K., Zhang, X., Ren, S., Sun, J.: Deep residual learning for image recognition. In: Proceedings of the IEEE Conference on Computer Vision and Pattern Recognition (CVPR), pp. 770–778 (2016)
6. Irvin, J., et al.: Chexpert: A large chest radiograph dataset with uncertainty labels and expert comparison. AAAI Press (2019)
7. Kasten-Serlin, S., Goldberger, J., Greenspan, H.: Adaptation of a multisite network to a new clinical site via batch-normalization similarity. In: The IEEE International Symposium on Biomedical Imaging (ISBI) (2022)
8. Kingma, D.P., Ba, J.: Adam: A method for stochastic optimization. arXiv preprint arXiv:1412.6980 (2014)
9. Krizhevsky, A., Sutskever, I., Hinton, G.E.: ImageNet classification with deep convolutional neural networks. In: Advances in Neural Information Processing Systems, pp. 1106–1114 (2012)
10. Lemaître, G., Martí, R., Freixenet, J., Vilanova, J.C., Walker, P.M., Meriaudeau, F.: Computer-aided detection and diagnosis for prostate cancer based on mono and multi-parametric MRI: a review. CBM **60**, 8–31 (2015)
11. Li, X., Grandvalet, Y., Davoine, F.: Explicit inductive bias for transfer learning with convolutional networks. In: International Conference on Machine Learning (ICML) (2018)
12. Litjens, G., et al.: Evaluation of prostate segmentation algorithms for MRI: the PROMISE12 challenge. MIA **18**(2), 359–373 (2014)

13. Liu, Q., Dou, Q., Heng, P.-A.: Shape-aware meta-learning for generalizing prostate MRI segmentation to unseen domains. In: Martel, A.L. (ed.) MICCAI 2020. LNCS, vol. 12262, pp. 475–485. Springer, Cham (2020). https://doi.org/10.1007/978-3-030-59713-9_46
14. Nikolov, S., et al.: Deep learning to achieve clinically applicable segmentation of head and neck anatomy for radiotherapy. CoRR abs/1809.04430 (2018)
15. Sagie, N., Greenspan, H., Goldberger, J.: Transfer learning with a layer dependent regularization for medical image segmentation. In: Lian, C., Cao, X., Rekik, I., Xu, X., Yan, P. (eds.) MLMI 2021. LNCS, vol. 12966, pp. 161–170. Springer, Cham (2021). https://doi.org/10.1007/978-3-030-87589-3_17
16. Shirokikh, B., Zakazov, I., Chernyavskiy, A., Fedulova, I., Belyaev, M.: First U-Net layers contain more domain specific information than the last ones. In: Albarqouni, S. (ed.) DART/DCL -2020. LNCS, vol. 12444, pp. 117–126. Springer, Cham (2020). https://doi.org/10.1007/978-3-030-60548-3_12
17. Souza, R., et al.: An open, multi-vendor, multi-field-strength brain MR dataset and analysis of publicly available skull stripping methods agreement. Neuroimage **170**, 482–494 (2018)
18. Takayama, K., Sato, I., Suzuki, T., Kawakami, R., Uto, K., Shinoda, K.: Smooth transfer learning for source-to-target generalization. In: NeurIPS Workshop on Distribution Shifts: Connecting Methods and Applications (2021)
19. Tsai, Y.H., Hung, W.C., Schulter, S., Sohn, K., Yang, M.H., Chandraker, M.: Learning to adapt structured output space for semantic segmentation. In: Proceedings of the IEEE Conference on Computer Vision and Pattern Recognition (2018)
20. Wang, M., Deng, W.: Deep visual domain adaptation: a survey. Neurocomputing **312**, 135–153 (2018)
21. Wang, X., Peng, Y., Lu, L., Lu, Z., Bagheri, M., Summers, R.M.: ChestX-ray8: Hospital-scale chest x-ray database and benchmarks on weakly-supervised classification and localization of common thorax diseases. IEEE Computer Society (2017)
22. Yosinski, J., Clune, J., Bengio, Y., Lipson, H.: How transferable are features in deep neural networks? (2014)
23. Zakazov, I., Shirokikh, B., Chernyavskiy, A., Belyaev, M.: Anatomy of domain shift impact on U-Net layers in MRI segmentation. In: de Bruijne, M. (ed.) MICCAI 2021. LNCS, vol. 12903, pp. 211–220. Springer, Cham (2021). https://doi.org/10.1007/978-3-030-87199-4_20

Stain-AgLr: Stain Agnostic Learning for Computational Histopathology Using Domain Consistency and Stain Regeneration Loss

Geetank Raipuria, Anu Shrivastava, and Nitin Singhal(✉)

Advanced Technology Group, AIRA MATRIX, Mumbai, India
{geetank.raipuria,anu.shrivastava,nitin.singhal}@airamatrix.com

Abstract. Stain color variations between Whole Slide Images (WSIs) is a key challenge in the application of Computational Histopathology. Deep learning-based algorithms are susceptible to domain shift and degrade in performance on the WSIs captured from a different source than the training data due to stain color variations. We propose a training methodology Stain-AgLr, that achieves high invariance to stain color changes on unseen test data. In addition to task loss, Stain-AgLr training is supervised with a consistency regularization loss that enforces consistent predictions for training samples and their stain altered versions. An additional decoder is used to regenerate stain color from feature representation of the stain altered images. We compare the proposed approach to state-of-the-art strategies using two histopathology datasets and show significant improvement in model performance on unseen stain variations. We also visualize the feature space distribution of test samples from multiple diagnostic labs and show that Stain-AgLr achieves a significant overlap between the distributions.

Keywords: Stain invariance · Domain generalization · Histopathology

Deep Learning based computational histopathology has demonstrated the ability to enhance patient healthcare quality by automating the time-consuming and expensive task of analyzing high-resolution WSIs [4,9]. The analysis involves identifying tissue or cellular level morphological features highlighted by staining dyes like Hematoxylin and Eosin (H&E). The stain color distribution of a WSI depends upon factors including the process of tissue preparation, dye manufacturer, and scanning equipment. As a result, there exists a high variability in the appearance of histopathology images, as seen in Fig. 1. Stain variations may appear between WSIs scanned at different centers as well as within the same center [2]. Often the model training data is obtained from single lab, but the model is deployed across multiple other labs. This domain shift hampers the performance of deep learning models on out-of-distribution samples [11,22,29], as seen in Fig. 2a.

K. Kamnitsas et al. (Eds.): DART 2022, LNCS 13542, pp. 33–44, 2022.
https://doi.org/10.1007/978-3-031-16852-9_4

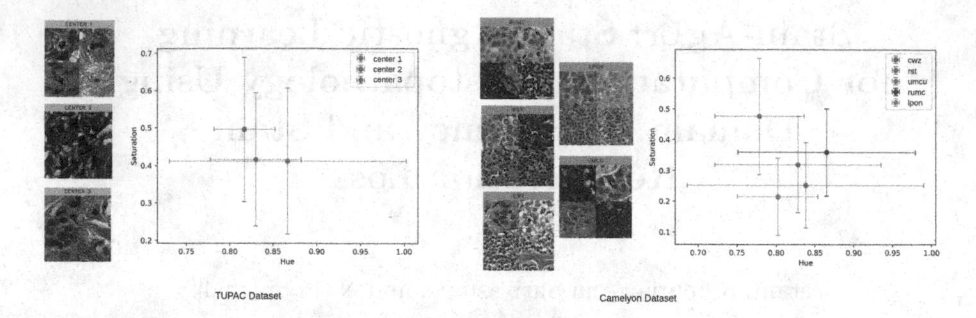

Fig. 1. Variation in stain color distribution between data from different labs. The graph shows the mean and standard deviation of pixel intensity for all image patches in the respective dataset, in the HSV color space. Best viewed in color.

Related Work: Existing solutions reduce the effect of variation in stain distribution by normalizing stain color or by improving the model's generalization on the test set. Traditional stain normalization approaches [17,26] normalize the target domain images by matching their color distribution to a single reference image from the source domain. Recently, generative deep CNN [3,8,13,19,20] have been trained to perform image-to-image domain transfer, learning the color distribution from the entire set of source domain images. However, the regenerated images may display undesired artifacts [3,7], leading to misdiagnosis. Furthermore, stain normalization significantly adds to the computation cost as each image needs to be pre-processed. Model generalization to stain variations can also be improved by learning stain agnostic features using Domain Adversarial training [12,15] or by using color augmentation to simulate variations in stain [6,10,25,27,28]. These methods provide better generalization without the need for a dedicated normalization network during inference.

Many state-of-the-art unsupervised approaches [3,6,16,27,28,31] rely on unlabeled images from the test set to generalize the model to a target distribution. [6,27,28] adapt the training data image stain based on samples from test lab data to use as augmentations, [3,20] require images from target domain to learn a stain normalization model, where as [16,31] use semi-supervised models for unsupervised domain adaptation, learning from unlabelled target domain images. Similarly, [31] showed that semi-supervised learning methods like [21,23,30] can learn from a consistency loss between real and a noisy version of the unlabeled target domain images. However, computational histopathology models need to be invariant to unseen intra-lab as well as inter-lab stain variations for application in patient care. In this work, we propose a novel strategy that learns stain invariant features without requiring any knowledge of the test data distribution. The goal of our approach is to learn a feature space that has a high overlap between the distribution of training and unseen test set with stain variations.

Our Contributions: Stain Invariance is induced with a two pronged strategy, as shown in Fig. 3. We use a stain altered version of an image to mimick test samples from a different lab, and impose consistency between the prediction of

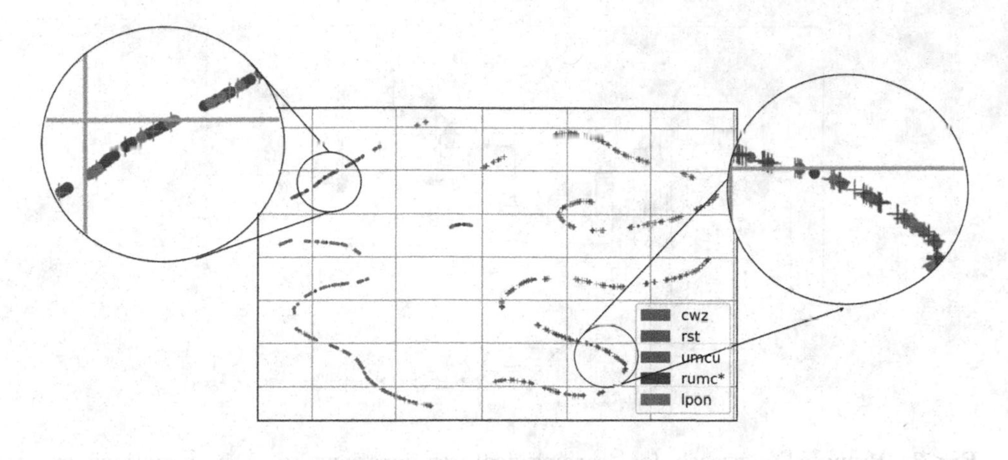

(a) Vanilla Model

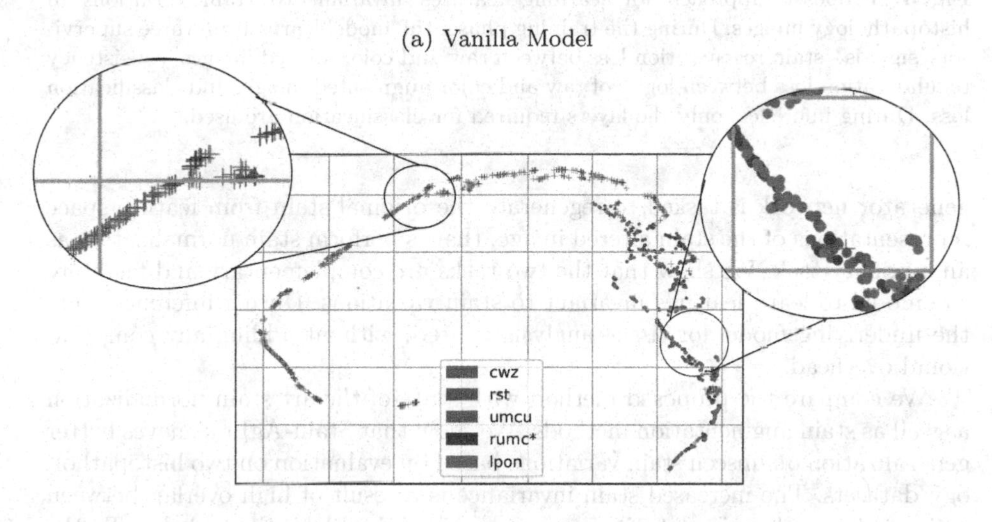

(b) Stain-AgLr

Fig. 2. The figure shows scatter plots of 2D features obtained using UMAP [18] for validation data from RUMC Lab and Test set data from CWZ, RST, UMCU and LPON labs for Camleyon [1] dataset. Symbols + and • represents class 0 and 1 respectively. For test set data, Vanilla model mis-classifies samples, represented by + over-lapping with • and vice-versa. Stain-AgLr model classifies a significantly smaller number of samples incorrectly, with only a few • over-lapping with +. This shows that the feature space produced by proposed Stain-AgLr model shows a high overlap between the validation and test distributions.

the raw image and its stain altered version, by penalizing their relative entropy. The consistency regularization loss enforces similar feature space representation for differently stained version of the same images, that is, the model learns that the difference in stain color has no bearing on the prediction task. In parallel, a

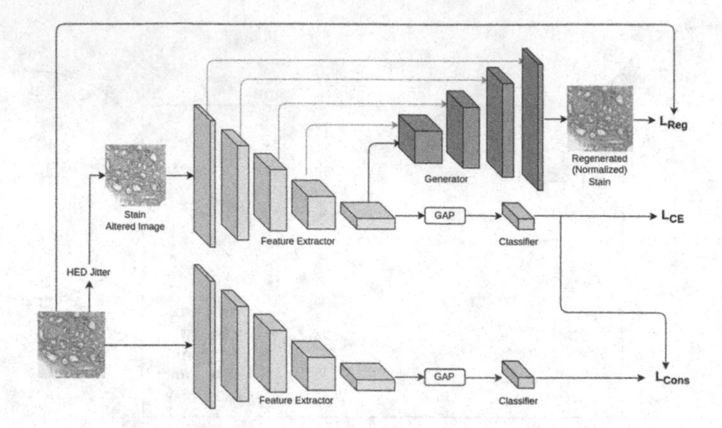

Fig. 3. Proposed approach for learning features invariant to stain variations in histopathology images. During the training phase, the model learns from three supervisory signals - stain regeneration loss between raw and color altered images, consistency regularization loss between logits of raw and color augmented image, and classification loss. During inference, only the layers required for classification are used.

generator network is tasked to regenerate the original stain from feature space representations of the stain altered image, that is perform stain normalization as an auxiliary task. We show that the two tasks are complementary and facilitate the model to learn features invariant to stain variations. During inference, only the underlying model for tissue analysis is used, without adding any computational overhead.

We compare the proposed method with state-of-the-art stain normalization as well as stain augmentation methods. We show that Stain-AgLr achieves better generalization on unseen stain variations based on evaluation on two histopathology datasets. The increased stain invariance is a result of high overlap between train and test domain data in feature space produced by Stain-AgLr. To the best of our knowledge, our work is the first to employ stain normalization as an auxiliary task rather than a preprocessing step and show that it leads to improved generalization on unseen test data with stain variations. Furthermore, the inference time corresponds to that of only the classification network which is significantly lower compared to stain normalization methods.

1 Method

We train a classification network that shares a feature extractor with a stain regeneration network. In addition to Cross-Entropy Loss, the network is supervised by two loss functions - Consistency Regularization Loss and Stain Regeneration Loss.

1.1 Model Architecture

Let M_{ft}, M_{cls} & M_{gen} represent the feature extractor, the classifier, and the generator respectively. Together, the networks M_{ft} and M_{gen} constitute a stain regeneration network that learns the mapping to regenerate the stain color distribution of the training images from stain altered images. On the other hand, the network M_{ft}, in conjunction with M_{cls}, classifies the input image into a set of task specific classes. Global average pooling (GAP) and Dropout (50%) is applied on the output of the M_{ft}, which generates a feature vector that is the input to M_{cls}. During inference, only the classification network is used, with layers particular to the stain regeneration network removed. We follow CNN architecture provided by [10,25]. Details are provided in Supplementary Material.

1.2 HED Jitter

We employ HED jitter to generate stain altered histopathology images [24,25]. This results in samples that resemble data from different sources. As shown in Fig. 3, the model is fed with both raw and HED jittered images in a single batch. The altered image is utilized to learn stain invariant features by matching logits of the altered image to the original image, as well as train the stain regeneration network which regenerates the raw image. We use HED-light [25] configuration for Stain-AgLr with default parameters, including morphological and brightness-contrast augmentations.

1.3 Loss Functions

Two loss functions - Consistency Regularization and Stain Regeneration are used to train the Stain-AgLr model, in addition to task specific Cross-Entropy Loss. We guide the model to produce similar predictions for a raw image and its stain altered version using the Consistency Regularization Loss (L_{Cons}). Specifically, we minimize the divergence $D(P_\theta(y|x)||P_\theta(y|x,\epsilon))$, where D is Kullback-Leibler (KL) divergence loss, y is the ground truth label corresponding to input x and ϵ represents stain color noise. This enforces the model to be insensitive to stain color noise.

The features of the altered image from M_{ft} are passed to the M_{gen} which generates a stain color to match the raw image. We use MSE loss as the Stain Regeneration Loss (L_{Reg}) between the raw image and the regenerated image. This auxiliary task helps the model improve generalization on images with stain variations, by learning shared features useful for both classification and regeneration tasks. As a result, a combination of three loss functions is used to train the model.

$$L = L_{CE} + \lambda_1 L_{Reg} + \lambda_2 L_{Cons} \tag{1}$$

$$L_{Cons} = D_{KL}\big(M_{cls}(M_{ft}(x)) \,||\, M_{cls}(M_{ft}(\hat{x}))\big) \tag{2}$$

$$L_{Reg} = \frac{1}{W*H} \sum_{i=1}^{W*H} ||M_{gen}(M_{ft}(\hat{x})) - x||^2 \quad (3)$$

where λ_1 and λ_2 are weights, x and $\hat{x}$ are the raw and stain altered images.

2 Experiments

2.1 Setup

We evaluate Stain-AgLr using two publically available datasets - TUPAC Mitosis Detection and Camelyon17 Tumor Metastasis Detection. Both datasets segregate images based on the lab of origin. This allows models to be trained on data from a single lab, while unseen data samples from other labs are utilized to test the model's robustness to stain variations.

Camelyon17. [1]. The dataset contains H&E stained WSI of sentinel lymph nodes from five different medical centers: In our experiments, 10 WSI are used from each center for which annotation masks are available. 95500 (256×256 size) patches were created at 40x magnification, of which 48300 represent metastasis. Patches from RUMC were used for training and validation, remaining centers (CWZ, UMCU, RST, LPON) are used as test sets.

TUPAC Mitosis Detection. [5]. The dataset consists of 73 breast cancer cases from three pathology centers. The first 23 cases were obtained from a single center, whereas cases 24–48 and 48–73 were collected from two other centers. We use binary labels provided by [5], which comprises of 1,898 Mitotic figures and 5,340 Hard Negative patches of size 128×128. For training and validation, we use samples from the first 23 cases and separately report performance on the other two subsets.

Training Setup: We use an initial learning rate of 5e-3 for TUPAC dataset and 1e-2 for Camelyon dataset, obtained using a grid search. In the case of TUPAC, we sample an equal number of images per batch for both classes to mitigate the effect of class imbalance. All models are trained using the Adam optimizer with a batch size of 64, reducing the learning rate by a factor of 0.1 if the validation loss does not improve for 4 epochs for Camelyon and 15 epochs for TUPAC. The training was stopped when the learning rate dropped to 1e-5. For each run, we select the model weights corresponding to the model with the lowest validation loss. We found re-weighing factors $\lambda_1 = 0.1$ & $\lambda_2 = 10$ in multi-task loss (Eq. 3) gave the best performance. Geometric augmentations including random rotation in multiples of 90°, horizontal and vertical flipping were employed for all models. HED-Light Augmentation is used with default parameters as described in [25], including morphological and brightness contrast augmentation. Models were trained using NVIDIA Tesla A100 GPUs using PyTorch Library.

2.2 Evaluation

We conduct experiments to compare our proposed method with stain normalization as well as color augmentation techniques for improving model generalisability on unseen test data. Tables 1 and 2 report AUC scores on test data obtained from multiple labs with differing stain color distributions, along with standard deviations. All experiments are repeated ten times, with different random seeds.

Table 1. Classification AUC scores on Camelyon17 Metastasis Detection dataset. The model is trained on data from RUMC lab, evaluated on test set from labs with unseen stain variations.

Method	CWZ	RST	UMCU	LPON	MEAN
# patches	13527	9690	26967	14867	65051
Vanilla	0.9247 ± 0.011	0.8013 ± 0.026	0.8509 ± 0.040	0.7518 ± 0.1057	0.8321
Vahadane	0.9810 ± 0.004	0.9457 ± 0.009	0.9029 ± 0.032	0.8022 ± 0.039	0.9079
STST GAN	**0.9923 ± 0.002**	0.9766 ± 0.004	0.9533 ± 0.006	0.8471 ± 0.015	0.9423
DSCSI GAN	0.9800 ± 0.004	0.9731 ± 0.006	0.9689 ± 0.007	0.9029 ± 0.015	0.9563
HED-Light Aug	0.9790 ± 0.005	0.9708 ± 0.002	0.9424 ± 0.010	0.9264 ± 0.012	0.9546
HED-Light Aug+L_{Reg}	0.9849 ± 0.001	0.9713 ± 0.003	0.9509 ± 0.009	0.9446 ± 0.007	0.9623
HED-Light Aug+L_{Cons}	0.9892 ± 0.002	0.9749 ± 0.005	0.9670 ± 0.010	0.9373 ± 0.009	0.9671
HED-Light Aug+L_{Reg} + L_{Cons} Stain-AgLr	0.9903 ± 0.001	**0.9784 ± 0.002**	**0.9733 ± 0.004**	**0.9474 ± 0.008**	**0.9724**

Vanilla model does not use any stain normalization or stain augmentation. *Vahadane* [26], *STST GAN*[19] and *DSCSI GAN*[13] represent classifier performance on normalized images using the corresponding stain normalization method. Both [13,19] do not require samples from target domain during training the GAN. *HED-Light Aug* represents a model trained with HED-Light augmentation. Lastly, *Stain-AgLr* represents the proposed approach.

Table 2. Classification AUC scores on TUPAC Mitosis Detection dataset. The models were trained on data from Center 1, and data from Center 2 & 3 is used as test set with unseen stain variations.

Method	Center 2	Center 3	Time in Sec
# patches	1600	1344	100
Vanilla	0.6899 ± 0.046	0.8439 ± 0.031	**0.29**
Vahadane	0.7299 ± 0.037	0.8520 ± 0.025	102.29
STST GAN	0.7405 ± 0.023	0.8596 ± 0.009	0.82
DCSCI GAN	0.7483 ± 0.037	0.8643 ± 0.017	0.91
HED-Light Aug	0.7527 ± 0.032	0.8657 ± 0.021	**0.29**
HED-Light Aug+L_{Reg}	0.7971 ± 0.015	0.8935 ± 0.014	**0.29**
HED-Light Aug+L_{Cons}	0.7719 ± 0.036	0.8916 ± 0.020	**0.29**
HED-Light Aug+L_{Reg} + L_{Cons} Stain-AgLr	**0.8026 ± 0.015**	**0.8996 ± 0.012**	**0.29**

3 Discussion

Classifiers trained on data from one lab show poor performance on data from other labs. The performance degradation depends upon the deviation of stain color distribution from the training set. All stain normalization methods improve the classifier performance, thus inducing invariance to stain color changes in the downstream classification model. Both GAN-based approaches [13,19] provide better stain normalization, learning stain color distribution from the entire training set unlike [26] that uses a single reference image from the training set. We observe that, a classifier trained with HED-light augmentation matches or out performs deep learning-based stain normalization approaches, as also reported by [22,25]. This indicates that the vanilla model overfits the color stain information from a single lab, which is alleviated by use of color augmentation.

Impact of Consistency Regularization and Stain Regeneration Loss. Stain-AgLr outperforms models trained using stain normalization algorithms as well as HED-light augmentation. The enhancement in stain invariance is contributed by both Consistency Regularization as well as the Stain Regeneration Loss. Using the loss individually improves model performance over model trained with HED-Light augmentation, however, best results are obtained by the network trained using both the loss functions together. This demonstrates that the two tasks are complementary to one another for learning stain invariant features. The multi-task network learns shared representation that is less likely to over-fit on the noise in the form of stain color.

Stain invariance of Stain-AgLr can be further established by analyzing the distribution of validation set and test data in feature space using UMAP plots, visualized in Fig. 2. For the test set data, the plots show significantly better class separation produced by the proposed Stain-AgLr model as compared to the Vanilla model. Importantly, the distribution of class samples from the test set corresponds with respective classes from validation data. In other words, the feature space produced by proposed Stain-AgLr model shows a high overlap between the validation and test data distributions. This verifies that quantitative performance gain is obtained by Stain-AgLr learning stain invariant features.

All stain normalization approaches significantly increase the base classifier's inference time significantly, as seen in Table 2. Although Stain-AgLr employs additional convolution layers during training, the model's inference time is identical to that of a vanilla classifier. A higher throughput is beneficial in reducing turnaround time in patient diagnostics, especially when processing high resolution Histopathology WSIs, as well as reducing computational requirement for deployment in diagnostics laboratory setup. Thus, the proposed approach combines the best of both worlds: improved stain invariance and fast inference.

4 Conclusion

Invariance to stain color variations in histopathology images is essential for the effective deployment of computational models. We present a novel technique -

Stain-AgLr, which learns stain invariant features that lead to improved performance on images from different labs. We also show that Stain-AgLr results in a high overlap between feature space distributions of images with varying H&E staining. Unlike many state-of-the-art techniques, Stain-AgLr does not require unlabelled images from the test data as well as does not add any computational burden during inference.

A Qualitative Results

See Fig. 4

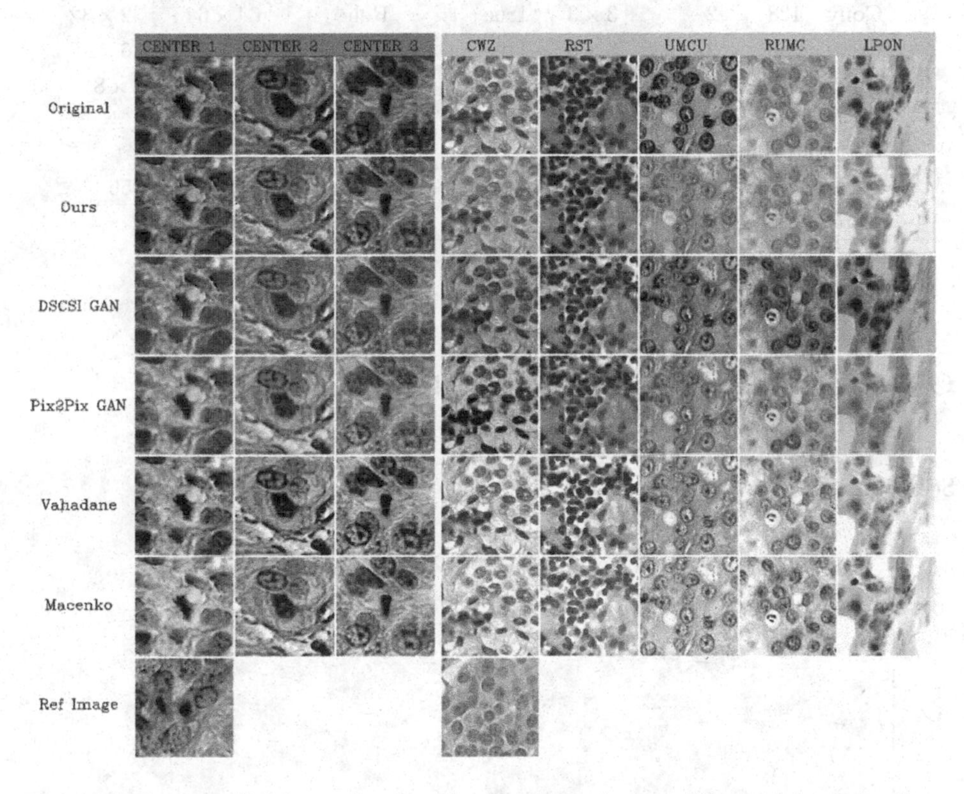

Fig. 4. Stain normalized samples for all subsets acquired at different labs. Data from Center 1 and RUMC is used as the training for TUPAC and CAMELYON respectively. Deep Learning based models, DCSCI GAN [13], STST GAN [19], and Stain-AgLr (Ours) learn stain color distribution from the entire training dataset; whereas traditional stain normalization methods including Vahadane [26] and Macenko [14] use a single image as reference, shown in the last row. Stain-CRS does not use stain normalized images, rather the normalized image is output of the stain regeneration auxiliary task.

B Model Architecture Details

See Table 3

Table 3. Description of CNN model architecture used for all experiments. We follow CNN architecture provided by [10,25]. Similarly, M_{gen} uses stacked convolutions in each stage before upsampling to get the original image size.

		Filters	# Conv	Kernel	Batch norm	Activation	Input	Output
M_{ft}	Conv	32	2	3×3	True	Relu	256×256	128×128
	Conv	64	2	3×3	True	Relu	128×128	64×64
	Conv	128	2	3×3	True	Relu	64×64	32×32
	Conv	256	2	3×3	True	Relu	32×32	16×16
	Conv	512	2	3×3	True	Relu	16×16	8×8
	Global Average Pooling Dropout (0.5)							
M_{cls}	Dense	256				Relu	512	256
	Dense	2				Relu	256	2

C Weightage of Consistency Regularization and Regeneration Loss

See Fig. 5

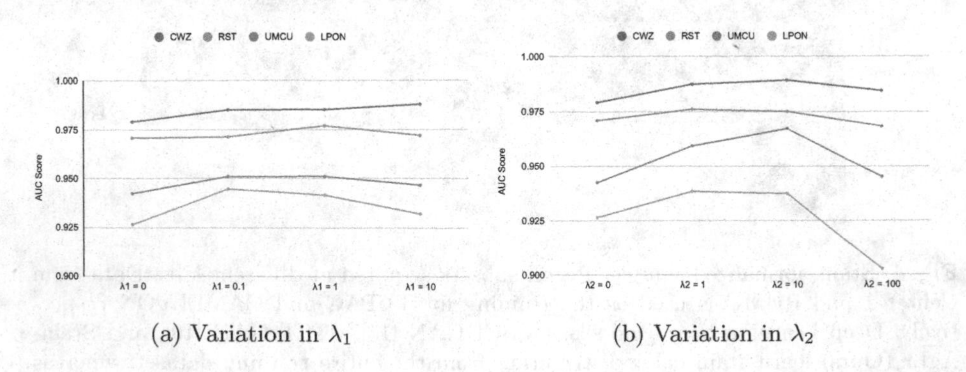

(a) Variation in λ_1 (b) Variation in λ_2

Fig. 5. Performance of the proposed model with different λ_1 and λ_2 values - weightage of Stain Regeneration and Consistency Regularization loss, keeping the other $\lambda = 0$, evaluated on Camelyon dataset

References

1. Bandi, P., et al.: From detection of individual metastases to classification of lymph node status at the patient level: the camelyon17 challenge. IEEE Trans. Med. Imaging **38**(2), 550–560 (2018)
2. Bejnordi, B.E., et al.: Stain specific standardization of whole-slide histopathological images. IEEE Trans. Med. Imaging **35**(2), 404–415 (2015)
3. de Bel, T., Bokhorst, J.M., van der Laak, J., Litjens, G.: Residual cyclegan for robust domain transformation of histopathological tissue slides. Med. Image Anal. **70**, 102004 (2021)
4. Bera, K., Schalper, K.A., Rimm, D.L., Velcheti, V., Madabhushi, A.: Artificial intelligence in digital pathology-new tools for diagnosis and precision oncology. Nat. Rev. Clin. Oncol. **16**(11), 703–715 (2019)
5. Bertram, C.A., et al.: Are pathologist-defined labels reproducible? comparison of the tupac16 mitotic figure dataset with an alternative set of labels. In: Cardoso, J., et al. (eds.) IMIMIC/MIL3ID/LABELS -2020. LNCS, vol. 12446, pp. 204–213. Springer, Cham (2020). https://doi.org/10.1007/978-3-030-61166-8_22
6. Chang, J.-R., et al.: Stain mix-up: Unsupervised domain generalization for histopathology images. In: de Chang, M., et al. (eds.) MICCAI 2021. LNCS, vol. 12903, pp. 117–126. Springer, Cham (2021). https://doi.org/10.1007/978-3-030-87199-4_11
7. Cohen, J.P., Luck, M., Honari, S.: Distribution matching losses can hallucinate features in medical image translation. In: Frangi, A.F., Schnabel, J.A., Davatzikos, C., Alberola-López, C., Fichtinger, G. (eds.) MICCAI 2018. LNCS, vol. 11070, pp. 529–536. Springer, Cham (2018). https://doi.org/10.1007/978-3-030-00928-1_60
8. Cong, C., Liu, S., Di Ieva, A., Pagnucco, M., Berkovsky, S., Song, Y.: Semi-supervised adversarial learning for stain normalisation in histopathology images. In: de Bruijne, M., et al. (eds.) MICCAI 2021. LNCS, vol. 12908, pp. 581–591. Springer, Cham (2021). https://doi.org/10.1007/978-3-030-87237-3_56
9. Cui, M., Zhang, D.Y.: Artificial intelligence and computational pathology. Lab. Invest. **101**(4), 412–422 (2021)
10. Faryna, K., van der Laak, J., Litjens, G.: Tailoring automated data augmentation to h&e-stained histopathology. In: Medical Imaging with Deep Learning (2021)
11. Koh, P.W., et al.: Wilds: a benchmark of in-the-wild distribution shifts. In: International Conference on Machine Learning, pp. 5637–5664. PMLR (2021)
12. Lafarge, M.W., Pluim, J.P.W., Eppenhof, K.A.J., Moeskops, P., Veta, M.: Domain-adversarial neural networks to address the appearance variability of histopathology images. In: Cardoso, M.J., et al. (eds.) DLMIA/ML-CDS -2017. LNCS, vol. 10553, pp. 83–91. Springer, Cham (2017). https://doi.org/10.1007/978-3-319-67558-9_10
13. Liang, H., Plataniotis, K.N., Li, X.: Stain style transfer of histopathology images via structure-preserved generative learning. In: Deeba, F., Johnson, P., Würfl, T., Ye, J.C. (eds.) MLMIR 2020. LNCS, vol. 12450, pp. 153–162. Springer, Cham (2020). https://doi.org/10.1007/978-3-030-61598-7_15
14. Macenko, M., et al.: A method for normalizing histology slides for quantitative analysis. In: 2009 IEEE International Symposium on Biomedical Imaging: From Nano to Macro, pp. 1107–1110. IEEE (2009)
15. Marini, N., Atzori, M., Otálora, S., Marchand-Maillet, S., Muller, H.: H&e-adversarial network: a convolutional neural network to learn stain-invariant features through hematoxylin & eosin regression. In: Proceedings of the IEEE/CVF International Conference on Computer Vision, pp. 601–610 (2021)

16. Melas-Kyriazi, L., Manrai, A.K.: Pixmatch: Unsupervised domain adaptation via pixelwise consistency training. In: Proceedings of the IEEE/CVF Conference on Computer Vision and Pattern Recognition, pp. 12435–12445 (2021)
17. Reinhard, E., Adhikhmin, M., Gooch, B., Shirley, P.: Color transfer between images. IEEE Comput. Graphics Appl. **21**(5), 34–41 (2001)
18. Sainburg, T., McInnes, L., Gentner, T.Q.: Parametric umap: learning embeddings with deep neural networks for representation and semi-supervised learning. ArXiv e-prints (2020)
19. Salehi, P., Chalechale, A.: Pix2pix-based stain-to-stain translation: A solution for robust stain normalization in histopathology images analysis. In: 2020 International Conference on Machine Vision and Image Processing (MVIP), pp. 1–7. IEEE (2020)
20. Shaban, M.T., Baur, C., Navab, N., Albarqouni, S.: Staingan: Stain style transfer for digital histological images. In: 2019 IEEE 16th international symposium on biomedical imaging (Isbi 2019), pp. 953–956. IEEE (2019)
21. Sohn, K., et al.: Fixmatch: simplifying semi-supervised learning with consistency and confidence. Adv. Neural. Inf. Process. Syst. **33**, 596–608 (2020)
22. Stacke, K., Eilertsen, G., Unger, J., Lundström, C.: A closer look at domain shift for deep learning in histopathology. arXiv preprint arXiv:1909.11575 (2019)
23. Tarvainen, A., Valpola, H.: Mean teachers are better role models: Weight-averaged consistency targets improve semi-supervised deep learning results. In: Advances in Neural Information Processing Systems, vol. 30 (2017)
24. Tellez, D., et al.: Whole-slide mitosis detection in h&e breast histology using phh3 as a reference to train distilled stain-invariant convolutional networks. IEEE Trans. Med. Imaging **37**(9), 2126–2136 (2018)
25. Quantifying the effects of data augmentation and stain color normalization in convolutional neural networks for computational pathology. Med. Image Anal. **58**, 101544 (2019)
26. Vahadane, A., et al.: Structure-preserving color normalization and sparse stain separation for histological images. IEEE Trans. Med. Imaging **35**(8), 1962–1971 (2016)
27. Vasiljević, J., Feuerhake, F., Wemmert, C., Lampert, T.: Towards histopathological stain invariance by unsupervised domain augmentation using generative adversarial networks. Neurocomputing **460**, 277–291 (2021)
28. Wagner, S.J., et al.: Structure-Preserving Multi-domain Stain Color Augmentation Using Style-Transfer with Disentangled Representations. In: de Bruijne, M., et al. (eds.) MICCAI 2021. LNCS, vol. 12908, pp. 257–266. Springer, Cham (2021). https://doi.org/10.1007/978-3-030-87237-3_25
29. Wang, M., Deng, W.: Deep visual domain adaptation: a survey. Neurocomputing **312**, 135–153 (2018)
30. Xie, Q., Dai, Z., Hovy, E., Luong, T., Le, Q.: Unsupervised data augmentation for consistency training. Adv. Neural. Inf. Process. Syst. **33**, 6256–6268 (2020)
31. Zhang, Y., Zhang, H., Deng, B., Li, S., Jia, K., Zhang, L.: Semi-supervised models are strong unsupervised domain adaptation learners. arXiv preprint arXiv:2106.00417 (2021)

MetaMedSeg: Volumetric Meta-learning for Few-Shot Organ Segmentation

Azade Farshad[1(✉)], Anastasia Makarevich[1], Vasileios Belagiannis[2], and Nassir Navab[1,3]

[1] Technical University of Munich, Munich, Germany
azade.farshad@tum.de
[2] Otto von Guericke University Magdeburg, Magdeburg, Germany
[3] Johns Hopkins University, Baltimore, USA

Abstract. The lack of sufficient annotated image data is a common issue in medical image segmentation. For some organs and densities, the annotation may be scarce, leading to poor model training convergence, while other organs have plenty of annotated data. In this work, we present MetaMedSeg, a gradient-based meta-learning algorithm that redefines the meta-learning task for the volumetric medical data with the goal of capturing the variety between the slices. We also explore different weighting schemes for gradients aggregation, arguing that different tasks might have different complexity and hence, contribute differently to the initialization. We propose an importance-aware weighting scheme to train our model. In the experiments, we evaluate our method on the medical decathlon dataset by extracting 2D slices from CT and MRI volumes of different organs and performing semantic segmentation. The results show that our proposed volumetric task definition leads to up to 30% improvement in terms of IoU compared to related baselines. The proposed update rule is also shown to improve the performance for complex scenarios where the data distribution of the target organ is very different from the source organs. (Project page: http://metamedseg.github.io/)

1 Introduction

Segmentation of medical images is an effective way to assist medical professionals in their diagnosis. Recent advances in deep learning have made it possible to achieve high accuracy in organ and tumour segmentation [18,24]. Despite the recent advances in medical image segmentation, standard supervised learning settings usually require a large amount of labelled data. Labelled data can be abundantly available for some organs (e.g., liver), yet it can be very scarce for others.

A. Farshad and A. Makarevich—Contributed equally to this work.

Supplementary Information The online version contains supplementary material available at https://doi.org/10.1007/978-3-031-16852-9_5.

K. Kamnitsas et al. (Eds.): DART 2022, LNCS 13542, pp. 45–55, 2022.
https://doi.org/10.1007/978-3-031-16852-9_5

One of the early approaches to overcome this limitation is transfer learning, where a neural network is pertained on a large labelled dataset (source domain) and then fine-tuned on a small amount of labelled data (target domain) [23]. In such scenarios, domain adaptation and generalization [13,17] is of high interest. Another common approach that gained much popularity is few-shot learning which aims to learn from just a few examples. One recent example of few-shot learning is COVID-19 detection using chest X-rays [11].

Few-shot methods can be roughly divided into augmentation-based learning and task-based meta-learning. In this work we focus on the meta-learning approach, which also comes in different flavours: metric learning (e.g., prototypical networks [29,34,35]), memory-based learning [10,33] and gradient-based methods. Few-shot learning for image segmentation has been an active research topic [2,4,9,16,27,31,32], but there are few works [7,19,21,25] focusing on meta-learning [14,15], zero-shot [6] and few-shot medical image segmentation.

In this work, we adopt a meta-learning approach, relying on one of the recent modifications of the MAML (Model-Agnostic Meta-Learning) algorithm [8] - Reptile [20], which is a simple and yet effective algorithm for meta-training a model and adapt it to our problem. We address the two main components of gradient-based meta-learning: task definition and gradient aggregation. The tasks are defined by sampling pairs of images and their corresponding segmentation maps based on the specified criteria. We propose a volume-based task definition designed explicitly for volumetric data and introduce a weighting mechanism for aggregating the gradients in each meta-training step, beneficial to non-IID data (independent and identically distributed).

The main contribution of this work is the volumetric task definition. We show that sampling data from one volume per task could lead to better optimization of the local models on the specific organ due to more control over the shots' variability. In contrast to the standard setting, where tasks are sampled randomly and can as well end up with images from similar parts of the volume (e.g., just first slices of different volumes), we ensure that a certain level of diversity exists between the shots, but at the same time, the diversity of the source set in general is reduced; this is done due to the fact that each volume is associated with a fixed, limited number of shots (e.g., 15), and other shots never participate in the training, which has shown to have a positive effect on training, as shown in [26].

Our second contribution is the importance-aware weighting scheme. In the classical Reptile setting, gradients of sampled tasks are averaged in each meta-epoch, while in our proposed method, the gradients are weighted based on the importance of each task. The importance is defined as the distance between a local model trained on a given task and the average of all models trained on other tasks. This weighting mechanism has been proposed in the federated learning framework [36] for non-IID data. We argue that giving less weight to tasks with a higher distance from the average model gives less chance to the outlier data, which could help avoid catastrophic forgetting (the tendency of the neural network to forget previously learnt information) [3]. This could also perform as a regularization to avoid overfitting when tasks are similar and benefit the training when the cross-domain distance is high.

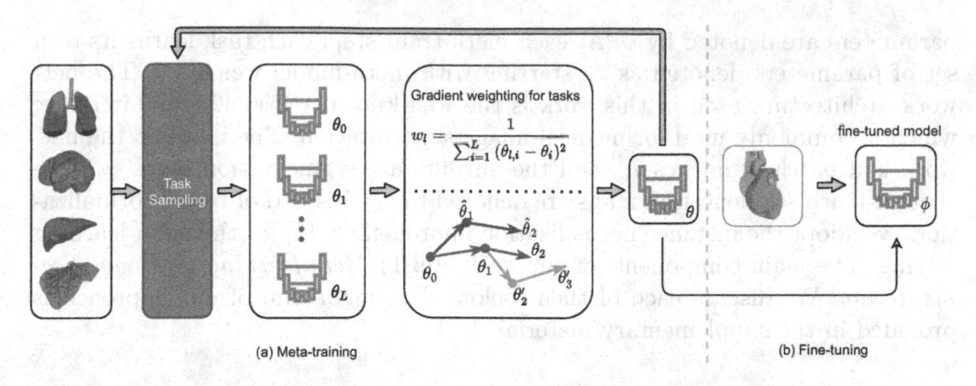

Fig. 1. Defining the meta-training tasks and weighting these tasks based on their importance is done in the meta-training step as demonstrated on the left. θ and ϕ' are the meta-model parameters, and the model in the progress of meta-training. The meta-trained model with parameters ϕ is then used for the final fine-tuning step on the target organ.

Our evaluations show that the proposed volumetric task definition and weighted gradient aggregation improve the segmentation's accuracy. We evaluate our method in two different settings: 1. Few-shot setting, where the models are fine-tuned on few shots. 2. Full-data, where the model is fine-tuned on all of the data for the target organ. We compare our results with multiple baselines: 1. Supervised learning with random initialization, 2. Supervised learning with transfer learning initialization, 3. Average task weighting baseline based on Reptile [20], 4. Few-shot cell segmentation by Dawoud et al. [7], which is the closest related work. We employ our proposed volumetric task definition on the baselines to show its effectiveness in different meta-learning settings. Figure 1 shows an overview of our method.

To summarize, we propose MetaMedSeg, a meta-learning approach for medical image segmentation. The main contributions of this work are as follows: 1. A novel task definition based on data volumes designed for medical scenarios 2. A novel update rule for few-shot learning where the cross-domain distance is high. 3. Significant improvement of segmentation performance compared to standard methods. The source code of this work will be publicly released upon its acceptance.

2 Methodology

Given a dataset $\mathcal{D} = \{\mathcal{S}, \mathcal{T}\}$, we define $\mathcal{S} = \{\mathcal{S}_1 \cdots \mathcal{S}_n\}$ as the training set (source domain) and $\mathcal{T} = \{\mathcal{T}_1 \cdots \mathcal{T}_m\}$ as the test set (target domain), where n, m are the number of organ datasets for the source and target domains and $\mathcal{T} \cap \mathcal{S} = \emptyset$. Each dataset consists of pairs of images and segmentation masks. The task in our setup is defined then as a subset of k shots sampled from $\mathcal{S}_i$ or $\mathcal{T}_j$.

The learning has two steps: meta-training and fine-tuning. The model parameters from the meta-training step are denoted by θ, while the fine-tuned model parameters are denoted by ϕ. At each meta-train step, each task learns its own set of parameters denoted as θ_l, starting with meta-model weights θ. The network architecture used in this work is the well-known U-Net [24] architecture, which is commonly used for medical image segmentation. The input to the network is a batch of images $\mathcal{I}_b$, and the outputs are segmentation maps y_b. The ground-truth segmentation maps are denoted by y'_b. Instead of batch normalization, we adopt the instance normalization approach [5,12] for the meta-learning setting. The main components of our work are: 1) *Meta-learning*, 2) *Image Segmentation.* We discuss each of them below. The algorithm of our approach is provided in the supplementary material.

2.1 Meta-learning

As shown in Fig. 1, in each round of meta-training, a set of tasks consisting of images and their corresponding segmentation maps are sampled according to the chosen rule and trained with the U-Net model. The trained models' learnt weights are then aggregated based on the specified update rule. The final model is used as initialization for the fine-tuning step. In classical Reptile algorithm [20], the updates obtained from all tasks are averaged in each meta-epoch. We propose a different strategy for weighting these updates based on the importance of the tasks. The details of task definition, task sampling, and update rules will be discussed next.

Task Definition. A task in meta-learning for segmentation can be defined in different ways. Our initial approach is based on [7]. A task is a set of k images and masks belonging to the same dataset. For example, k-shots sampled randomly from all available vessel cancer slices is such a task. This approach is targeted mainly at 2D data, and although we are working with 2D slices, some meta-information from the 3D data could be used. Our data is not just a set of images but a set of volumes (3D tensors that can be sliced along chosen direction to produce sets of 2D images), so we propose a volume-based task definition. We suggest defining a *task* as a set of images from the same volume $\mathcal{V}$ sampled with the step size $= \left\lceil \frac{|\mathcal{V}|}{K} \right\rceil$; this ensures the balancedness of task sizes across datasets.

Weighted Task Sampling. Some organs have different modalities or different zones (e.g., the prostate can be split into peripheral and transitional zones), which we treat as separate datasets. One can see that this might lead to some organs dominating during task sampling (e.g., the BRATS dataset alone is translated into 12 different source datasets). To counter that, we suggest using weighted sampling to give each organ a fair chance to get into the tasks set in each meta-epoch. For each organ with z different modalities or zones, we set the sampling rate of each modality/zone to $\frac{1}{z}$ and normalize them.

Importance-Aware Task Weighting. We employ the original Reptile [20] update rule as a baseline for our method:

$$\theta \leftarrow \theta + \beta \frac{1}{L} \sum_{l=1}^{L} (\theta_l - \theta), \tag{1}$$

where θ_l is the weights vector of the local model, θ is the weights vector of the meta-model, L is the number of tasks, and β is the meta learning rate. We compare the following settings:

- Average weighting (AW): all updates have the same weights, and $(\theta_l - \theta)$ are averaged across tasks.
- Inverse distance weighting (IDW): more weight is given to models closer to the meta-model.

The weights for AW update are defined by the number of tasks sampled in each meta-epoch. With L tasks sampled, the weights for each of tasks would be: $\frac{1}{L}$. The weighting for the inverse distance update is given by:

$$w_l = \frac{1}{\sum_{i=1}^{L} (\theta_{l,i} - \theta_i)^2}, \tag{2}$$

where θ_i is the i-th weight of the meta-model and $w_{l,i}$ is the i-th weight of the l-th task's model of the current meta-epoch. The weights are also normalized to sum up to 1 using $w_l = \frac{w_l}{\sum_j^L w_j}$. The update rule is therefore:

$$\theta \leftarrow \theta + \beta \sum_{l=1}^{L} w_l (\theta_l - \theta), \tag{3}$$

2.2 Image Segmentation

We employ the weighted BCE loss in combination with the approximation of Intersection over Union (IoU) loss [22] for segmentation. The details of loss functions and the derivations are reported in the supplementary material.

$$L(y, y') = BCE(y, y') - \log\left(\frac{2IoU(y, y')}{IoU(y, y') + 1}\right) \tag{4}$$

3 Experiments

We train and evaluate our approach on the medical decathlon dataset [28], which consists of 3D MRI and CT volumes of 9 organs, namely the brain (368 volumes), hippocampus (260), lung (25), prostate (32), cardiac (20), pancreas (279), colon (121), hepatic vessels (216), and spleen (41). We use the U-Net [24] architecture and train our model using IoU, BCE, and the combination of the two losses.

We implement the transfer learning approach as a baseline, where we train with all available training data to obtain the initialization weights for fine-tuning. To ensure that the results obtained in the experiments are not due to the specific k-shot selection, we perform fine-tuning on 5 random selections and compute the average IoU over those runs tested on the same test set of previously unseen data.

3.1 Experimental Setup

The data is pre-processed by splitting different techniques (e.g., T2 and FLAIR) and different regions (e.g., edema and tumor) into separate datasets, which results in 24 different datasets. For each dataset, we set a threshold indicating whether we consider an object present in the image or not based on the number of pixels and visual inspection of the results. The values of thresholds can be found in the supplementary material.

All the images were resized to 256×256 resolution. The threshold was applied after resizing. We also apply volume normalization during slicing by subtracting the mean and dividing by the standard deviation of all the non-zero pixels of the whole volume. We fix the same conditions for all fine-tuning experiments: we train for 20 epochs using weight decay 3×10^{-5}, learning rate $\alpha = 0.005$ with a step learning rate decay of $\gamma = 0.7$ at every other step. For full-data training, we use a learning rate 0.001 and weight decay $w = 3 \times 10^{-5}$.

For meta-training, we train for 100 epochs, sampling 5 tasks with 15 shots and 1 image per shot at each meta-epoch. We used a learning rate of $\alpha = 0.01$ for local and meta-model with weight decay of $w = 0.003$ and the same learning rate decay described above. For transfer learning, we train on all datasets excluding cardiac, prostate, and spleen for 20 epochs. For hyperparameter-tuning, we used the particle swarm optimization approach. No augmentations were performed on the data in any of the settings.

3.2 Results and Discussion

Table 1 shows the comparison of our work to the baseline methods in the few-shot setting for 15 shots on four different organs, as well as The performance of the models fine-tuned on the whole support set. Some qualitative examples of segmentation results are shown in Fig. 2. The results of our experiments show that volume-based task design has the most effect on the segmentation performance, especially in the full-data setting. The effect of our proposed update rule is minimal in some cases but visible in other organs, such as Prostate Transitional; this could be due to the nature of data distribution and the shapes of organs. When organ shapes have a high diversity (e.g., prostate compared to other organs), the outlier shapes could benefit from the proposed inverse distance update rule, which gives less weight to gradients further from average.

Table 1. Comparison of our proposed methods to related work in full data and few-shot setting on 4 different organs. AW and IDW stand for Average weighting and Inverse distance weighting respectively. Prostate Peripheral and Prostate Transitional are denoted by Prostate P. and Prostate T. respectively. Vol. refers to volumetric task.

Update rule	Target organ	15-shots - IoU ↑		Full data - IoU ↑	
		Standard	Vol. (ours)	Standard	Vol. (ours)
Supervised learning	Cardiac	58.78	–	90.26	–
Transfer learning	Cardiac	66.22	–	90.46	–
Dawoud et al. [7]	Cardiac	68.28	67.7	90.38	92.47
MetaMedSeg + IDW (ours)	Cardiac	67.49	64.86	91.38	94.51
MetaMedSeg + AW (ours)	Cardiac	67.83	**68.33**	91.08	**95.55**
Supervised learning	Spleen	38.81	–	86.74	–
Transfer learning	Spleen	51.18	–	86.10	–
Dawoud et al. [7]	Spleen	49.29	**58.34**	89.65	87.53
MetaMedSeg + IDW (ours)	Spleen	55.64	50.50	89.96	91.53
MetaMedSeg + AW (ours)	Spleen	55.98	56.44	90.00	**92.27**
Supervised Learning	Prostate P.	7.35	–	39.06	–
Transfer Learning	Prostate P.	10.87	–	39.94	–
Dawoud et al. [7]	Prostate P.	15.99	12.82	37.05	41.58
MetaMedSeg + IDW (ours)	Prostate P.	17.15	13.89	47.58	68.26
MetaMedSeg + AW (ours)	Prostate P.	16.17	**22.69**	46.20	**70.90**
Supervised learning	Prostate T.	38.64	–	68.04	–
Transfer learning	Prostate T.	41.09	–	69.84	–
Dawoud et al. [7]	Prostate T.	42.85	46.28	70.42	71.55
MetaMedSeg + IDW (ours)	Prostate T.	44.25	44.72	68.98	**79.95**
MetaMedSeg + AW (ours)	Prostate T	42.43	**48.33**	67.68	78.84

We hypothesize that, when using volumetric task definition, all organs have the chance to contribute to the final model in a more balanced setting. Therefore, the performance decreases when the weighted update rule is combined with volumetric tasks. Another reason could be that the weighting of the updates helps in biasing the model towards tasks that are more similar to the target. To understand the effect of image diversity, we create an adjacency matrix of the average Euclidean distances between pairs of randomly selected volumes. The results in Fig. 3 show that in our target organs, the following have the lowest to highest distance: cardiac, spleen, and prostate. The effect of this distance is visible in the final segmentation performance of each organ.

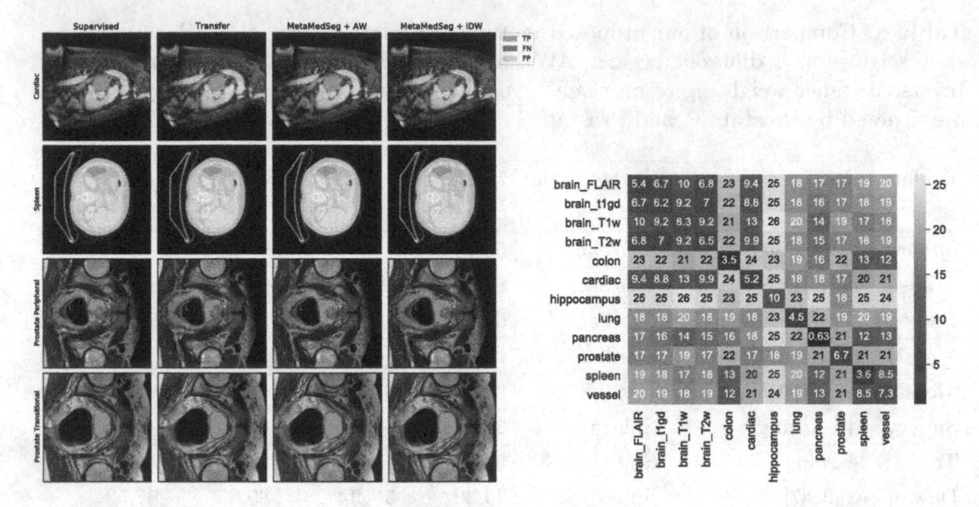

Fig. 2. A comparison of different segmentation baselines with our method for four different target organs in the full-data setting.

Fig. 3. Heatmap of the average distances between pairs of images from different organs

Table 2. Ablation study of our method with different losses in few-shot setting for cardiac segmentation. AW, IDW and Vol. stand for Average weighting, Inverse distance weighting and volumetric task definition respectively.

Update rule	Segmentation loss	IoU ↑	
		Standard	Vol. (ours)
AW	IoU	67.82	68.13
IDW	IoU	67.42	64.86
AW	Tversky Focal loss	65.90	65.72
IDW	Tversky Focal loss	63.71	62.78
AW	Dice loss	65.87	66.03
IDW	Dice loss	62.58	62.73
AW	BCE	**67.83**	**68.33**
IDW	BCE	65.09	64.29
AW	BCE + IoU	66.85	67.30
IDW	BCE + IoU	66.98	64.71

3.3 Ablation Study

We show the effect of the proposed update rule, different segmentation losses, and the volume-based task definition used in this work in Table 2. The models were meta-trained using five different losses, including Focal Tversky loss [1] and Dice loss [30], and then fine-tuned using IoU loss. The best performance is achieved using weighted BCE loss for meta-training and IoU loss for fine-tuning.

4 Conclusion

We presented a novel way of task definition for few-shot learning for volume-based 2D data and an update rule based on the importance of tasks in the meta-training step. Our method is evaluated on four different organ types with the least amount of data, namely cardiac, spleen, prostate peripheral, and prostate transitional. Our approach applies not only to organ segmentation but also to tumour segmentation or other types of densities. The results show that our proposed volumetric task definition improves the segmentation performance significantly in all organs. The proposed update rules provide considerable improvement in terms of IoU. Both proposed approaches (volumetric tasks and weighted update rule) could be useful in different scenarios. While the volumetric task definition proves to be advantageous in all scenarios, it is more beneficial to use the weighted update rule when the data distribution of the target class is different from the source, for example in cases such as segmentation of new diseases where also the amount of labeled data is limited.

Acknowledgements. We gratefully acknowledge the Munich Center for Machine Learning (MCML) with funding from the Bundesministerium für Bildung und Forschung (BMBF) under the project 01IS18036B.

References

1. Abraham, N., Khan, N.M.: A novel focal tversky loss function with improved attention u-net for lesion segmentation. In: 2019 IEEE 16th International Symposium on Biomedical Imaging (ISBI 2019), pp. 683–687. IEEE (2019)
2. Azad, R., Fayjie, A.R., Kauffmann, C., Ben Ayed, I., Pedersoli, M., Dolz, J.: On the texture bias for few-shot cnn segmentation. In: Proceedings of the IEEE/CVF Winter Conference on Applications of Computer Vision, pp. 2674–2683 (2021)
3. Bertugli, A., Vincenzi, S., Calderara, S., Passerini, A.: Few-shot unsupervised continual learning through meta-examples. arXiv preprint arXiv:2009.08107 (2020)
4. Boudiaf, M., Kervadec, H., Masud, Z.I., Piantanida, P., Ben Ayed, I., Dolz, J.: Few-shot segmentation without meta-learning: A good transductive inference is all you need? In: Proceedings of the IEEE/CVF Conference on Computer Vision and Pattern Recognition, pp. 13979–13988 (2021)
5. Bronskill, J., Gordon, J., Requeima, J., Nowozin, S., Turner, R.: Tasknorm: rethinking batch normalization for meta-learning. In: International Conference on Machine Learning, pp. 1153–1164. PMLR (2020)
6. Chartsias, A., et al.: Multimodal cardiac segmentation using disentangled representation learning. In: Pop, M., et al. (eds.) STACOM 2019. LNCS, vol. 12009, pp. 128–137. Springer, Cham (2020). https://doi.org/10.1007/978-3-030-39074-7_14
7. Dawoud, Y., Hornauer, J., Carneiro, G., Belagiannis, V.: Few-shot microscopy image cell segmentation. In: Dong, Y., Ifrim, G., Mladenić, D., Saunders, C., Van Hoecke, S. (eds.) ECML PKDD 2020. LNCS (LNAI), vol. 12461, pp. 139–154. Springer, Cham (2021). https://doi.org/10.1007/978-3-030-67670-4_9
8. Finn, C., Abbeel, P., Levine, S.: Model-agnostic meta-learning for fast adaptation of deep networks. In: International Conference on Machine Learning, pp. 1126–1135. PMLR (2017)

9. Gairola, S., Hemani, M., Chopra, A., Krishnamurthy, B.: Simpropnet: improved similarity propagation for few-shot image segmentation. In: Bessiere, C. (ed.) Proceedings of the Twenty-Ninth International Joint Conference on Artificial Intelligence, IJCAI 2020 [scheduled for July 2020, Yokohama, Japan, postponed due to the Corona pandemic], pp. 573–579. ijcai.org (2020)
10. Hu, T., Yang, P., Zhang, C., Yu, G., Mu, Y., Snoek, C.G.: Attention-based multi-context guiding for few-shot semantic segmentation. In: Proceedings of the AAAI Conference on Artificial Intelligence, vol. 33, pp. 8441–8448 (2019)
11. Jadon, S.: Covid-19 detection from scarce chest x-ray image data using few-shot deep learning approach. In: Medical Imaging 2021: Imaging Informatics for Healthcare, Research, and Applications, vol. 11601, p. 116010X. International Society for Optics and Photonics (2021)
12. Jia, S., Chen, D.J., Chen, H.T.: Instance-level meta normalization. In: Proceedings of the IEEE/CVF Conference on Computer Vision and Pattern Recognition, pp. 4865–4873 (2019)
13. Khandelwal, P., Yushkevich, P.: A few-shot meta learning framework for domain generalization in medical imaging. In: Albarqouni, S., et al. (eds.) DART/DCL -2020. LNCS, vol. 12444, pp. 73–84. Springer, Cham (2020). https://doi.org/10.1007/978-3-030-60548-3_8
14. Li, X., Yu, L., Jin, Y., Fu, C.-W., Xing, L., Heng, P.-A.: Difficulty-aware meta-learning for rare disease diagnosis. In: Martel, A.L., et al. (eds.) MICCAI 2020. LNCS, vol. 12261, pp. 357–366. Springer, Cham (2020). https://doi.org/10.1007/978-3-030-59710-8_35
15. Liu, Q., Dou, Q., Heng, P.-A.: Shape-aware meta-learning for generalizing prostate mri segmentation to unseen domains. In: Martel, A.L., et al. (eds.) MICCAI 2020. LNCS, vol. 12262, pp. 475–485. Springer, Cham (2020). https://doi.org/10.1007/978-3-030-59713-9_46
16. Liu, W., Zhang, C., Lin, G., Liu, F.: Crnet: Cross-reference networks for few-shot segmentation. In: Proceedings of the IEEE/CVF Conference on Computer Vision and Pattern Recognition, pp. 4165–4173 (2020)
17. Liu, X., Thermos, S., O'Neil, A., Tsaftaris, S.A.: Semi-supervised meta-learning with disentanglement for domain-generalised medical image segmentation. In: de Semi-supervised meta-Bruijne, M., et al. (eds.) MICCAI 2021. LNCS, vol. 12902, pp. 307–317. Springer, Cham (2021). https://doi.org/10.1007/978-3-030-87196-3_29
18. Milletari, F., Navab, N., Ahmadi, S.A.: V-net: Fully convolutional neural networks for volumetric medical image segmentation. In: 2016 fourth international conference on 3D vision (3DV), pp. 565–571. IEEE (2016)
19. Mondal, A.K., Dolz, J., Desrosiers, C.: Few-shot 3d multi-modal medical image segmentation using generative adversarial learning. arXiv preprint arXiv:1810.12241 (2018)
20. Nichol, A., Achiam, J., Schulman, J.: On first-order meta-learning algorithms. arXiv preprint arXiv:1803.02999 (2018)
21. Ouyang, C., Biffi, C., Chen, C., Kart, T., Qiu, H., Rueckert, D.: Self-supervision with superpixels: training few-shot medical image segmentation without annotation. In: Vedaldi, A., Bischof, H., Brox, T., Frahm, J.-M. (eds.) ECCV 2020. LNCS, vol. 12374, pp. 762–780. Springer, Cham (2020). https://doi.org/10.1007/978-3-030-58526-6_45
22. Rahman, M.A., Wang, Y.: Optimizing intersection-over-union in deep neural networks for image segmentatio. In: Bebis, G., Bebis, G., et al. (eds.) ISVC 2016.

LNCS, vol. 10072, pp. 234–244. Springer, Cham (2016). https://doi.org/10.1007/978-3-319-50835-1_22
23. Rohrbach, M., Ebert, S., Schiele, B.: Transfer learning in a transductive setting. Adv. Neural. Inf. Process. Syst. **26**, 46–54 (2013)
24. Ronneberger, O., Fischer, P., Brox, T.: U-net: Convolutional networks for biomedical image segmentation. In: Navab, N., Hornegger, J., Wells, W.M., Frangi, A.F. (eds.) MICCAI 2015. LNCS, vol. 9351, pp. 234–241. Springer, Cham (2015). https://doi.org/10.1007/978-3-319-24574-4_28
25. Roy, A.G., Siddiqui, S., Pölsterl, S., Navab, N., Wachinger, C.: 'squeeze & excite'guided few-shot segmentation of volumetric images. Med. Image Anal. **59**, 101587 (2020)
26. Setlur, A., Li, O., Smith, V.: Is support set diversity necessary for meta-learning? arXiv preprint arXiv:2011.14048 (2020)
27. Siam, M., Oreshkin, B.N., Jagersand, M.: Amp: Adaptive masked proxies for few-shot segmentation. In: Proceedings of the IEEE/CVF International Conference on Computer Vision, pp. 5249–5258 (2019)
28. Simpson, A.L., et al.: A large annotated medical image dataset for the development and evaluation of segmentation algorithms. arXiv preprint arXiv:1902.09063 (2019)
29. Snell, J., Swersky, K., Zemel, R.: Prototypical networks for few-shot learning. In: Guyon, I., et al. (eds.) Advances in Neural Information Processing Systems, vol. 30. Curran Associates, Inc. (2017)
30. Sudre, C.H., Li, W., Vercauteren, T., Ourselin, S., Jorge Cardoso, M.: Generalised dice overlap as a deep learning loss function for highly unbalanced segmentations. In: Cardoso, M.J., et al. (eds.) DLMIA/ML-CDS -2017. LNCS, vol. 10553, pp. 240–248. Springer, Cham (2017). https://doi.org/10.1007/978-3-319-67558-9_28
31. Tian, P., Wu, Z., Qi, L., Wang, L., Shi, Y., Gao, Y.: Differentiable meta-learning model for few-shot semantic segmentation. In: Proceedings of the AAAI Conference on Artificial Intelligence, vol. 34, pp. 12087–12094 (2020)
32. Tian, Z., Zhao, H., Shu, M., Yang, Z., Li, R., Jia, J.: Prior guided feature enrichment network for few-shot segmentation. In: IEEE Annals of the History of Computing, pp. 1–1 (2020)
33. Wang, H., Zhang, X., Hu, Y., Yang, Y., Cao, X., Zhen, X.: Few-shot semantic segmentation with democratic attention networks. In: Vedaldi, A., Bischof, H., Brox, T., Frahm, J.-M. (eds.) ECCV 2020. LNCS, vol. 12358, pp. 730–746. Springer, Cham (2020). https://doi.org/10.1007/978-3-030-58601-0_43
34. Wang, K., Liew, J.H., Zou, Y., Zhou, D., Feng, J.: Panet: Few-shot image semantic segmentation with prototype alignment. In: Proceedings of the IEEE/CVF International Conference on Computer Vision, pp. 9197–9206 (2019)
35. Yang, B., Liu, C., Li, B., Jiao, J., Ye, Q.: Prototype mixture models for few-shot semantic segmentation. In: Vedaldi, A., Bischof, H., Brox, T., Frahm, J.-M. (eds.) ECCV 2020. LNCS, vol. 12353, pp. 763–778. Springer, Cham (2020). https://doi.org/10.1007/978-3-030-58598-3_45
36. Yeganeh, Y., Farshad, A., Navab, N., Albarqouni, S.: Inverse distance aggregation for federated learning with non-iid data. In: Albarqouni, S., et al. (eds.) DART/DCL -2020. LNCS, vol. 12444, pp. 150–159. Springer, Cham (2020). https://doi.org/10.1007/978-3-030-60548-3_15

Unsupervised Site Adaptation by Intra-site Variability Alignment

Shaya Goodman[1], Shira Kasten Serlin[1], Hayit Greenspan[1], and Jacob Goldberger[2](✉)

[1] Tel-Aviv University, Tel-Aviv, Israel
{yishayahug,kastenserlin}@mail.tau.ac.il, hayit@eng.tau.ac.il
[2] Bar-Ilan University, Ramat-Gan, Israel
jacob.goldberger@biu.ac.il

Abstract. A medical imaging network that was trained on a particular source domain usually suffers significant performance degradation when transferred to a different target domain. This is known as the domain-shift problem. In this study, we propose a general method for transfer knowledge from a source site with labeled data to a target site where only unlabeled data is available. We leverage the variability that is often present within each site, the *intra-site variability*, and propose an *unsupervised site adaptation* method that jointly aligns the intra-site data variability in the source and target sites while training the network on the labeled source site data. We applied our method to several medical MRI image segmentation tasks and show that it consistently outperforms state-of-the-art methods.

Keywords: Unsupervised domain adaptation · UDA · Intra-site variability · MRI segmentation

1 Introduction

Neural networks have been successfully applied to medical image analysis. Unfortunately, a model that is trained to achieve high performance on a certain dataset, often drops in performance when tested on medical images from different acquisition protocols or different clinical sites. This model robustness problem, known as domain shift, especially occurs in Magnetic Resonance Imaging (MRI) since different scanning protocols result in significant variations in slice thickness and overall image intensities. Site adaptation improves model generalization capabilities in the target site by mitigating the domain shift between the sites. Unsupervised Domain Adaptation (UDA) assumes the availability of data from the new site but without manual annotations. The goal of UDA is to train a network using both the labeled source site data and the unlabeled target site data to make accurate predictions about the target site data. In this study we

This research was supported by the Ministry of Science & Technology, Israel.
The first two authors contributed equally.

K. Kamnitsas et al. (Eds.): DART 2022, LNCS 13542, pp. 56–65, 2022.
https://doi.org/10.1007/978-3-031-16852-9_6

concentrate on segmentation tasks (see an updated review on UDA for segmentation in [6]). Another setup is supervised domain adaptation where we also have labeled data from the target site (see e.g. [10]).

Recent UDA methods include feature alignment adversarial networks that are based on learning domain-invariant features using a domain discriminator which is co-trained with the network [23,26,28]. Image alignment adversarial networks (e.g. [4,5,8]) translate the appearance from one domain to another using multiple discriminators and a pixel-wise cycle consistency loss. Seg-JDOT [1] solves a site adaptation scenario using optimal transport theory by presenting a domain shift minimization in the feature space. Li et al. Another approach is transferring the trained model to a new domain by modulating the statistics in the Batch-Normalization layer [13,16]. Some methods such as [2,12] suggest test-time adaptation methods.

Intra-site variability can result from multiple reasons in the medical space, including slice variability across an imaged organ, varying scanning protocols and differences in the patient population being imaged. The intra-variability of the data collected from the source and targets site is often based on similar factors. Importantly, this can be exploited in the site adaptation process. Recent studies on UDA for classification have used intra-site variability induced by different classes to divide the feature space into different subsets. [7,9,20,27]. Pseudo labels, which are produced for samples in the target domain, are used for domain alignment. These methods cannot be applied to segmentation tasks, as mentioned in [1], since the number of possible segmentation maps is exponentially larger than the number of classes in a classification task.

This gap has motivated us to look for a different approach for solving the domain shift problem for segmentation tasks. We present a domain adaptation approach that tackles the inter-domain shift by aligning the intra-variability of the source and target sites. Our approach consistently out-performs the state-of-the-art site adaptation methods on several publicly available medical images segmentation tasks. The code to reproduce our experiments is available at https://github.com/yishayahu/AIVA.git.

2 Site Adaptation Based on Intra-site Variability Alignment

We present an *unsupervised* site adaptation method that explicitly takes the intra-site variability into account. We concentrate on MRI image segmentation task. In this scenario, we are given a U-net network that was trained on the source site. We jointly align the feature space of the target site to the source site, so as when optimizing the model on the source site, we obtain a model that performs well on the target site as well. More specifically, our method minimizes the domain shift between the source and the target by aligning the intra-site variability of the target site with the intra-site variability of the source site. The intra-site variability is modeled by separately clustering the source and the target sites in a suitable embedded space. The centers of the clusters of the

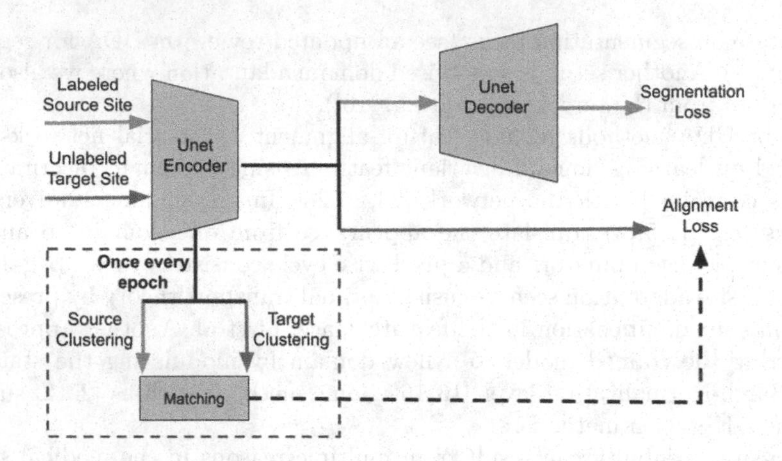

Fig. 1. A scheme of the fine-tuning loss assembly of the AIVA site adaptation method.

two sites are then matched, and each target cluster is pushed in towards its corresponding source cluster. In parallel, the segmentation loss is minimized on the source labeled data to maintain accurate semantic segmentation masks for the source site. Aligning the structure of the target site with the source site while maintaining good results on the source site, yields a good segmentation performance on the target site. In what follows, we provide a detailed description of each step of the proposed site adaptation algorithm.

Intra-site Variability Modeling. The intra-site variability is modeled by clustering the images of each site in a suitable embedded space. We compute an image embedding by considering the segmentation U-net bottleneck layer with its spatial dimensions and its convolutional filter dimension. We denote this image representation as the BottleNeck Space (BNS). Next, we apply the k-means algorithm to cluster the source site images in the BNS into k centers and in a similar manner we cluster the target site images into k centers. It is well known that applying k-means clustering to high-dimensional data does not work well because of the curse of dimensionality. Hence, in practice the actual clustering of the image representations is computed in a 2D embedding obtained by the PCA algorithm [11] followed by the t-SNE algorithm [19] that are applied jointly to the BNS representations of the source and target data points. We denote the 2D k clustering centers of the source site by $\{\mu_i^s\}_{i=1}^k$, and the 2D target site centers by $\{\mu_i^t\}_{i=1}^k$.

Clustering Matching. In this step, we align the intra-site variability structure of the target site to the source site by matching the two clusterings. We look for the optimal matching between the k source centers $\mu_1^s, ..., \mu_k^s$ and the k target centers $\mu_1^t, ..., \mu_k^t$:

$$\hat{\pi} = \arg\min_{\pi} \sum_{i=1}^{k} \|\mu_i^t - \mu_{\pi(i)}^s\|^2 \tag{1}$$

where π goes over all the $k!$ permutations. The Kuhn-Munkers matching algorithm, also known as the Hungarian method [14,21] is an algorithm that can efficiently solve the minimization problem (1) in time complexity $O(k^3)$. The clustering of the source and target images and the matching between the clusterings' centers are done once every epoch and are kept fixed throughout all the mini-batches of the epoch. This implies that the t-SNE procedure, the clustering and the matching algorithms do not need to be differentiable with respect to the model parameters since this process is separate from the backwards calculation of gradients and their impact on the total training running time is negligible (less than 2% addition to training time). Note that we can view the source (and target) site centers as the modes of a multi-modal distribution of the source (and target) data. Aligning the centers thus corresponds to aligning the source and target multi-modal distributions.

Alignment Loss. The assignment (1) found above is used to align the two sites by encouraging each target cluster center to be closer to the corresponding source center. Since in practice we work in mini-batches, we encourage the BNS representation of the average of target images in the current minibatch which were assigned to the same cluster, to be closer to the center of the corresponding source cluster. We define the following loss function in the BNS space:

$$L_{\text{alignment}} = \sum_{i=1}^{k} \|\bar{x}_i^t - \nu_{\hat{\pi}(i)}^s\|^2 \tag{2}$$

such that $\bar{x}_i^t$ is the average of all the target-site points in the minibatch that were assigned by the clustering procedure to the i-th cluster. The vector ν_i^s is the average of all source points that were assigned to the i-th cluster (μ_i^s is the average of the same set in the t-SNE embedded space). The domain shift between the source and target sites is thus minimized by aligning the data structure of the target site with the data structure of the source site. Note that in the alignment loss (2), while the source centers are kept fixed during an epoch, the target samples are obtained as a function of the model parameters, and the loss gradients with respect to the parameters are back propagated through them.

In addition to the alignment loss, we use a standard segmentation cross-entropy loss which is computed at the final output layer for the source samples and is designed to avoid degradation of the segmentation performances. Indirectly, it improves the segmentation of the target site data. The overall loss function is thus:

$$L = L_{\text{segmentation}} + \lambda L_{\text{alignment}}. \tag{3}$$

The regularization coefficient λ is a hyper-parameter that is usually tuned using cross-validation. Since there are no labels from the current target site, we cannot tune λ on a validation set. Instead, we use the following unsupervised tuning procedure: we average the values of $L_{\text{alignment}}$ in the first minibatches and define lambda as the reciprocal of the average. This makes the scaled alignment score close to 1 and makes it the same scale as our segmentation loss. The network is

Algorithm 1. The AIVA site adaptation algorithm

input: Source labeled images, target unlabeled images and a U-net that was trained on the source site.
for each epoch **do**
  Compute the BNS representation of all the source and the target samples.
  Cluster each site in a joint t-SNE embedding into k centers $\{\mu_i^s\}_{i=1}^k, \{\mu_i^t\}_{i=1}^k$.
  Find the optimal matching $\hat{\pi}$ between the source and the target centers:

$$\hat{\pi} = \arg\min_{\pi} \sum_{i=1}^{k} \|\mu_i^t - \mu_{\pi(i)}^s\|^2.$$

  for each mini-batch **do**
    For each cluster i, let $\bar{x}_i^t$ be the average of the target-site points (in the BNS domain) in the minibatch that were assigned to cluster i.
    Compute $L_{\text{alignment}} = \sum_{i=1}^{k} \|\bar{x}_i^t - \nu_{\hat{\pi}(i)}^s\|^2$ where ν_i^s is the average of all source-site points (in the BNS domain) that were assigned cluster i.
  Apply a gradient descent step to the loss: $L = L_{\text{segmentation}} + \lambda L_{\text{alignment}}$.
output: A model adapted to the target site.

pretrained on the source site, and then is adapted to the target site by minimizing the loss function (3). We dub the proposed method Adaptation by Intra-site Variability Alignment (AIVA). A scheme of the loss function of the AIVA algorithm is shown in Fig. 1. The AIVA algorithm is summarized in Algorithm Box 1.

3 Experiments

We evaluated the performance of our method and compared it with other unsupervised domain adaptation methods on two different medical image datasets for segmentation tasks. Our experiments were conducted on the following unsupervised domain adaptation setup: we have labeled data from a source site and unlabeled data from the target site and we are given a network that was trained on the source site data.

We chose a representative baseline from each of the three most dominant approaches today that deal with UDA (image statistics, domain shift minimization in feature space and feature alignment adversarial networks).

- *AdaBN*: recalculating the statistics of the batch normalization layers on the target site [16].
- *Seg-JDOT*: aligning the distributions of the source and the target sites using an optimal transport algorithm [1].
- *AdaptSegNet*: aligning feature space using adversarial learning [26].

We also directly trained a network on the target site using the labels of the training data of the target site, thereby setting an *upper bound* for UDA methods. In addition, we show the results on the pretrained model without any adaptation to set a lower bound.

MRI Skull Stripping: The publicly available dataset CC359 [25] consists of 359 MR images of heads where the task consists of skull stripping. The dataset

Table 1. Segmentation surface-Dice results on the brain MRI dataset CC359 [25].

Target site	Target site model	Source site model	Unsupervised adaptation			
			AdaBN	Seg-JDOT	AdaptSegNet	AIVA
Siemens, 1.5T	80.13	58.44	62.20	64.27	63.41	**67.43**
Siemens, 3T	80.19	59.65	58.82	61.83	57.57	**66.31**
GE, 1.5T	81.58	38.96	58.04	50.06	57.71	**59.95**
GE, 3T	84.16	56.27	54.23	59.82	55.94	**65.04**
Philips, 1.5T	84.00	56.38	73.01	69.59	68.22	**75.68**
Philips, 3T	82.19	41.93	51.18	50.30	50.51	**54.34**
Average	82.04	51.94	59.58	59.31	58.89	**64.79**

was collected from six sites which exhibit domain shift resulting in a severe score deterioration [24]. For preprocessing we interpolated to $1 \times 1 \times 1$ mm voxel spacing and scaled the intensities to a range of 0 to 1. To evaluate the different approaches, we used the surface Dice score [22] at a tolerance of 1 mm. While preserving consistency with the methodology in [24], we also found that surface Dice score to be a more suitable metric for the brain segmentation task than the standard Dice Score (similar to [29]). We used a U-net network that processes each 2D image slice separately. All the models were pretrained on a single source data for 5K steps starting with a learning rate of 10^{-3} that polynomially decays with an exponential power of 0.9 and a batch size 16. All compared models were finetuned using 6.5K steps. For AIVA we used 12 clusters. We ensured that all the models reached the loss plateau. Each target site was split into a training set and a test set. Since the assumption here was that we only has unlabeled images from the target site we chose the checkpoint using the performance on the source test set. We used 25 pairs of source and target sites and averaged the results of each target site. The remaining five pairs were used to examine the robustness of the method to different amount of clusters. The surface-Dice results are shown at Table 1. It highlights the significant deterioration between the supervised and the no-adaptation. Furthermore, we observe that our model consistently outperformed the baselines for each new site.

We visualize the alignment process in the AVIA algorithm. Intuitively we expect the intra-variability to be represented by the different clusters and the matching to align them across the source and the target. This is demonstrated in Fig. 2 (clusters 1–4) by examples from each cluster. Figure 3 shows the clustering of the source and target slices and the matching between the clusters. The two clusterings are similar, but not perfectly aligned due to the domain shift. Figure 4 shows that after the adaptation process the two sites are better aligned as a result of minimization of the alignment loss. Finally, Fig. 5 shows the sDice score as a function of the number of clusters (averaged over 5 source-target pairs). We can see that AIVA is robust to the amount of clusters when it is at least 9.

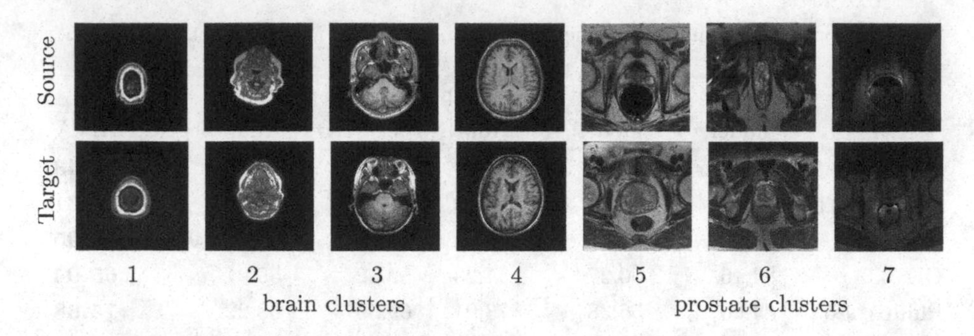

Fig. 2. Matching clusters images examples from source (top) and target (bottom) from CC359 brain dataset (1–4) and the prostate dataset (5–7).

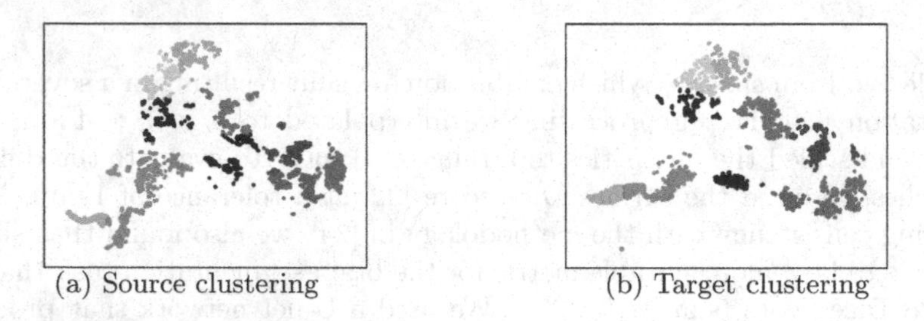

Fig. 3. Clustering of the slice samples of source (a) and target (b), at the beginning of the fine-tuning phase, in 2D space. Matched clusters - same color.

Prostate MRI Segmentation: To show the robustness of our method we evaluated it on a multi-source single-target setup as well. We used a publicly available multi-site dataset for prostate MRI segmentation which contains prostate T2-weighted MRI data (with segmentation masks) collected from different data sources with a distribution shift. Details of data and imaging protocols from the six different sites appear in [18]. Samples of sites A and B were taken from the NCI-ISBI13 dataset [3], samples of site C were from the I2CVB dataset [15], and samples of sites D, E and F were from the PROMISE12 dataset [17].

For pre-processing, we normalized each sample to have a zero mean and a unit variance in intensity value before inputting to the network. For each target site we used the other five sites as the source. The results were calculated on six possible targets. To evaluate different approaches, we used the Dice Score. We used the same network architecture and learning rate as in the experiment described above. We pretrained the network for 3.5K steps and finetuned the model for every method for another 3.5K steps. We ensured that all the models reached the loss plateau. Each site was split into a training set and a test set. We chose the checkpoint to evaluate using the source test set. We showed in the previous experiment that the AIVA algorithm is robust to the number of clusters. We fixed the number of clusters here to twelve as before.

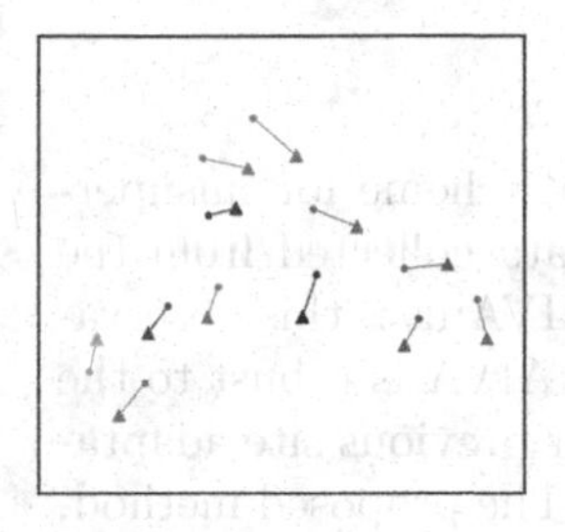

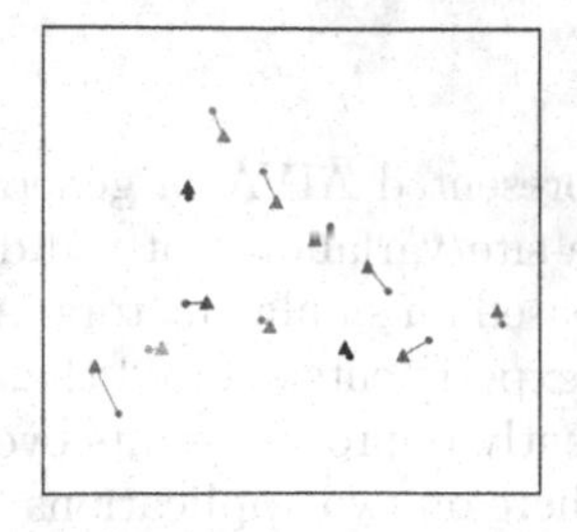

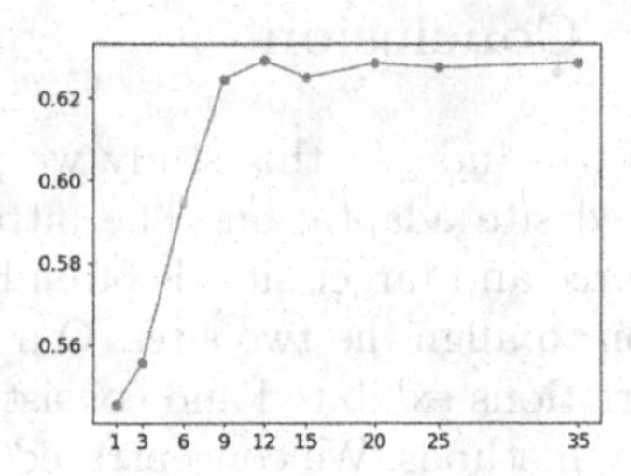

Fig. 4. Clusters' centers of the source (circles) and the target (triangles) in the 2D space before (left) and after (right) the adaptation phase.

Fig. 5. The AIVA sDice-score as a function of the of number of clusters.

Results. Figure 2 (5–7) shows the matching clusters in the training process: whereas in the Brain data the clusters focused on morphological variations, here we see a focus on the image contrast variability. In Table 2 we present comparative performances for each target site. We could not get a convergence for seg-JDOT [1] on this dataset, probably due to lack of data. Therefore, we omitted it from the result report. We note that AIVA yielded the overall best Dice score. In some sites, the difference between the supervised training and the source model is relatively small. For these cases, relatively weak results were seen for some of the UDA methods. AIVA showed stability by consistently yielding improved results. Examples of segmentation results are shown in Fig. 6.

Table 2. Segmentation Dice results on the prostate MRI dataset [18].

Target site	Target site model	Source site model	Unsupervised adaptation		
			AdaBN	AdaptSegNet	AIVA
Site A	87.48	**80.63**	79.03	76.63	79.94
Site B	85.62	59.84	73.32	64.45	**78.50**
Site C	78.86	58.84	68.08	67.64	**69.67**
Site D	85.30	61.78	66.99	77.39	**78.88**
Site E	80.55	77.72	**80.48**	79.13	79.65
Site F	86.10	79.98	53.36	80.88	**85.42**
Average	83.98	69.80	70.21	74.35	**78.67**

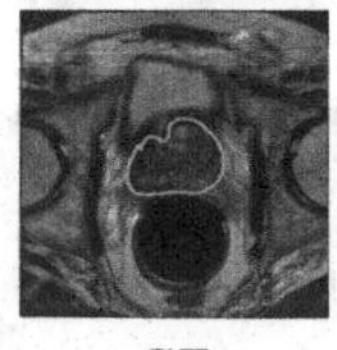

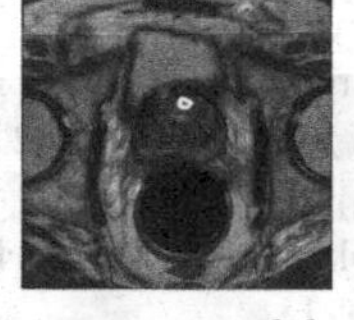

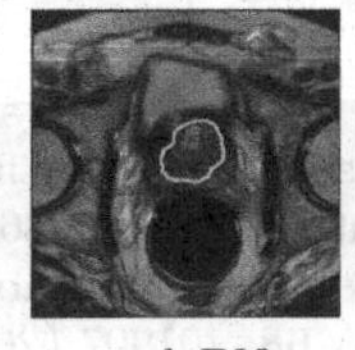

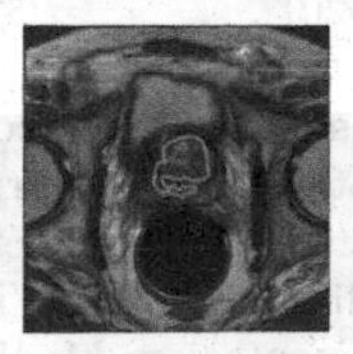

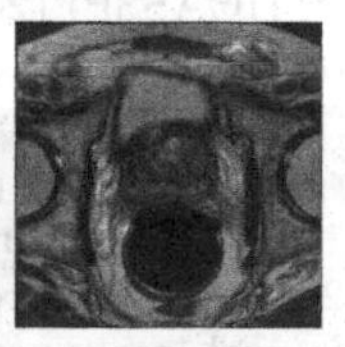

GT source model adaBN AdaptSegNet AIVA

Fig. 6. Qualitative segmentation results from the prostate MRI dataset.

4 Conclusion

To conclude, in this study we presented AIVA, a general scheme for unsupervised site adaptation. The intra-site variability of the data collected from the source and target sites is often based on similar factors. AIVA uses this observation to align the two sites. Our experiments showed that AIVA is robust to the variations exhibited and consistently improves results over previous site adaptation methods. We concentrated here on two applications. The proposed method, however, is general and is especially suitable for segmentation tasks where we cannot align the source and target site using the labels.

References

1. Ackaouy, A., Courty, N., Vallée, E., Commowick, O., Barillot, C., Galassi, F.: Unsupervised domain adaptation with optimal transport in multi-site segmentation of multiple sclerosis lesions from MRI data. Frontiers Comput. Neurosci. **14**, 19 (2020)
2. Bateson, M., Kervadec, H., Dolz, J., Lombaert, H., Ayed, I.B.: Source-relaxed domain adaptation for image segmentation (2020)
3. Bloch, N., et al.: NCI-ISBI 2013 challenge: automated segmentation of prostate structures. Cancer Imaging Arch. **370** (2015)
4. Chen, C., Dou, Q., Chen, H., Qin, J., Heng, P.A.: Synergistic image and feature adaptation: Towards cross-modality domain adaptation for medical image segmentation. In: Proceedings of the AAAI Conference on Artificial Intelligence (2019)
5. Chen, C., Dou, Q., Chen, H., Qin, J., Heng, P.A.: Unsupervised bidirectional cross-modality adaptation via deeply synergistic image and feature alignment for medical image segmentation. IEEE Trans. Med. Imaging **39**(7), 2494–2505 (2020)
6. Csurka, G., Volpi, R., Chidlovskii, B.: Unsupervised domain adaptation for semantic image segmentation: a comprehensive survey. arXiv preprint arXiv:2112.03241 (2021)
7. Deng, Z., Luo, Y., Zhu, J.: Cluster alignment with a teacher for unsupervised domain adaptation. In: International Conference on Computer Vision (ICCV) (2019)
8. Dou, Q., et al.: Pnp-adanet: Plug-and-play adversarial domain adaptation network with a benchmark at cross-modality cardiac segmentation. arXiv preprint arXiv:1812.07907 (2018)
9. Gao, B., Yang, Y., Gouk, H., Hospedales, T.M.: Deep clustering for domain adaptation. In: International Conference on Acoustics, Speech and Signal Processing (ICASSP) (2020)
10. Goodman, S., Greenspan, H., Goldberger, J.: Supervised domain adaptation using gradients transfer for improved medical image analysis. In: MICCAI Workshop on Domain Adaptation and Representation Transfer (DART) (2022)
11. Jolliffe, I.: Principal Component Analysis (1986)
12. Karani N, Erdil E, C.K.K.E.: Test-time adaptable neural networks for robust medical image segmentation. MedIA **68**, 101907 (2021)
13. Kasten-Serlin, S., Goldberger, J., Greenspan, H.: Adaptation of a multisite network to a new clinical site via batch-normalization similarity. In: The IEEE International Symposium on Biomedical Imaging (ISBI) (2022)

14. Kuhn, H.W.: The Hungarian method for the assignment problem. Naval Res. Logistics Q. **2**, 83–97 (1955)
15. Lemaître, G., Martí, R., Freixenet, J., Vilanova, J.C., Walker, P.M., Meriaudeau, F.: Computer-aided detection and diagnosis for prostate cancer based on mono and multi-parametric MRI: a review. CBM **60**, 8–31 (2015)
16. Li, Y., Wang, N., Shi, J., Hou, X., Liu, J.: Adaptive batch normalization for practical domain adaptation. Pattern Recogn. **80**, 109–117 (2018)
17. Litjens, G., et al.: Evaluation of prostate segmentation algorithms for MRI: the PROMISE12 challenge. MIA **18**(2), 359–373 (2014)
18. Liu, Q., Dou, Q., Heng, P.-A.: Shape-aware meta-learning for generalizing prostate MRI segmentation to unseen domains. In: Martel, A.L., et al. (eds.) MICCAI 2020. LNCS, vol. 12262, pp. 475–485. Springer, Cham (2020). https://doi.org/10.1007/978-3-030-59713-9_46
19. Van der Maaten, L., Hinton, G.: Visualizing data using t-SNE. J. Mach. Learn. Res. **9**(11), 2579–2605 (2008)
20. Menapace, W., Lathuilière, S., Ricci, E.: Learning to cluster under domain shift. In: European Conference on Computer Vision (2020)
21. Munkers, J.: Algorithms for the assignment and transportation problem. J. Soc. Ind. Appl. Math. **5**, 32–38 (1957)
22. Nikolov, S., et al.: Deep learning to achieve clinically applicable segmentation of head and neck anatomy for radiotherapy. CoRR abs/1809.04430 (2018)
23. Panfilov, E., Tiulpin, A., Klein, S., Nieminen, M.T., Saarakkala, S.: Improving robustness of deep learning based knee mri segmentation: Mixup and adversarial domain adaptation. In: Proceedings of the IEEE International Conference on Computer Vision Workshops (2019)
24. Shirokikh, B., Zakazov, I., Chernyavskiy, A., Fedulova, I., Belyaev, M.: First U-net layers contain more domain specific information than the last ones. In: Albarqouni, S., et al. (eds.) DART/DCL -2020. LNCS, vol. 12444, pp. 117–126. Springer, Cham (2020). https://doi.org/10.1007/978-3-030-60548-3_12
25. Souza, R., et al.: An open, multi-vendor, multi-field-strength brain MR dataset and analysis of publicly available skull stripping methods agreement. Neuroimage **170**, 482–494 (2018)
26. Tsai, Y.H., Hung, W.C., Schulter, S., Sohn, K., Yang, M.H., Chandraker, M.: Learning to adapt structured output space for semantic segmentation. In: Proceedings of the IEEE Conference on Computer Vision and Pattern Recognition (2018)
27. Xu, T., Chen, W., Wang, P., Wang, F., Li, H., Jin, R.: Cdtrans: Cross-domain transformer for unsupervised domain adaptation. In: International Conference on Learning Representations (ICLR) (2022)
28. Yan, W., Wang, Y., Xia, M., Tao, Q.: Edge-guided output adaptor: highly efficient adaptation module for cross-vendor medical image segmentation. IEEE Signal Process. Lett. **26**(11), 1593–1597 (2019)
29. Zakazov, I., Shirokikh, B., Chernyavskiy, A., Belyaev, M.: Anatomy of domain shift impact on U-net layers in MRI segmentation. In: de Bruijne, M., et al. (eds.) MICCAI 2021. LNCS, vol. 12903, pp. 211–220. Springer, Cham (2021). https://doi.org/10.1007/978-3-030-87199-4_20

Discriminative, Restorative, and Adversarial Learning: Stepwise Incremental Pretraining

Zuwei Guo[1], Nahid Ul Islam[1], Michael B. Gotway[2], and Jianming Liang[1](✉)

[1] Arizona State University, Tempe, AZ 85281, USA
{Zuwei.Guo,nuislam,jianming.liang}@asu.edu
[2] Mayo Clinic, Scottsdale, AZ 85259, USA
Gotway.Michael@mayo.edu

Abstract. Uniting three self-supervised learning (SSL) ingredients (discriminative, restorative, and adversarial learning) enables collaborative representation learning and yields three transferable components: a discriminative encoder, a restorative decoder, and an adversary encoder. To leverage this advantage, we have redesigned five prominent SSL methods, including Rotation, Jigsaw, Rubik's Cube, Deep Clustering, and TransVW, and formulated each in a *United* framework for 3D medical imaging. However, such a United framework increases model complexity and pretraining difficulty. To overcome this difficulty, we develop a stepwise incremental pretraining strategy, in which a discriminative encoder is first trained via discriminative learning, the pretrained discriminative encoder is then attached to a restorative decoder, forming a skip-connected encoder-decoder, for further joint discriminative and restorative learning, and finally, the pretrained encoder-decoder is associated with an adversarial encoder for final full discriminative, restorative, and adversarial learning. Our extensive experiments demonstrate that the stepwise incremental pretraining stabilizes United models training, resulting in significant performance gains and annotation cost reduction via transfer learning for five target tasks, encompassing both classification and segmentation, across diseases, organs, datasets, and modalities. This performance is attributed to the synergy of the three SSL ingredients in our United framework unleashed via stepwise incremental pretraining. All codes and pretrained models are available at GitHub.com/JLiangLab/StepwisePretraining.

Keywords: Self-supervised learning · Discriminative learning · Restorative learning · Adversarial learning · United framework · Stepwise pretraining

1 Introduction

Self-supervised learning (SSL) [11] pretrains generic source models [20] without using expert annotation, allowing the pretrained generic source models to be

---

Supplementary Information The online version contains supplementary material available at https://doi.org/10.1007/978-3-031-16852-9_7.

K. Kamnitsas et al. (Eds.): DART 2022, LNCS 13542, pp. 66–76, 2022.
https://doi.org/10.1007/978-3-031-16852-9_7

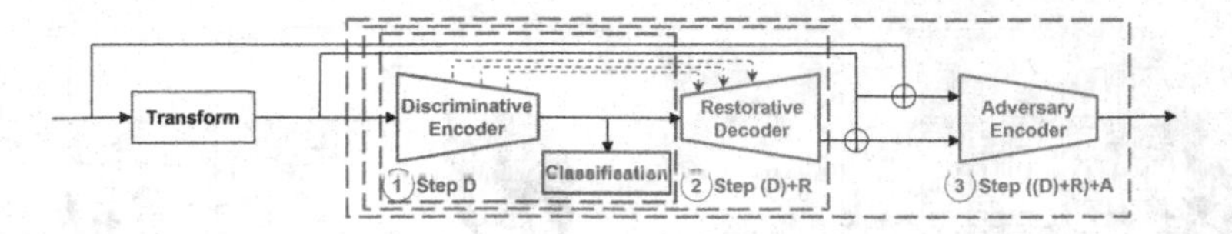

Fig. 1. Our United model consists of three components: a discriminative encoder, a restorative decoder, and an adversary encoder, where the discriminative encoder and the restorative decoder are skip connected, forming an encoder-decoder. To overcome the United model complexity and pretraining difficulty, we develop a strategy to incrementally train the three components in a stepwise fashion: (1) Step D trains the discriminative encoder via discriminative learning; (2) Step (D)+R attaches the pretrained discriminative encoder to the restorative decoder for further joint discriminative and restorative learning; (3) Step ((D)+R)+A associates the pretrained encoder-decoder with the adversarial encoder for final full discriminative, restorative, and adversarial learning. This stepwise incremental pretraining has proven to be reliable across multiple SSL methods (Fig. 2) for a variety of target tasks across diseases, organs, datasets, and modalities.

quickly fine-tuned into high-performance application-specific target models with minimal annotation cost [18]. The existing SSL methods may employ one or a combination of the following three learning ingredients [9]: (1) discriminative learning, which pretrains an encoder by distinguishing images associated with (computer-generated) pseudo labels; (2) restorative learning, which pretrains an encoder-decoder by reconstructing original images from their distorted versions; and (3) adversarial learning, which pretrains an additional adversary encoder to enhance restorative learning. Haghighi et al. articulated a vision and insights for integrating three learning ingredients in one single framework for collaborative learning [9], yielding three learned components: a discriminative encoder, a restorative decoder, and an adversary encoder (Fig. 1). However, such integration would inevitably increase model complexity and pretraining difficulty, raising these two questions: *(a) how to optimally pretrain such complex generic models and (b) how to effectively utilize pretrained components for target tasks?*

To answer these two questions, we have redesigned five prominent SSL methods for 3D imaging, including Rotation [7], Jigsaw [13], Rubik's Cube [21], Deep Clustering [4], and TransVW [8], and formulated each in a single framework called "United" (Fig. 2), as it unites discriminative, restorative, and adversarial learning. Pretraining United models, i.e., all three components together, directly from scratch is unstable; therefore, we have investigated various training strategies and discovered a stable solution: stepwise incremental pretraining. An example of such pretraining follows: first training a discriminative encoder via discriminative learning (called Step D), then attaching the pretrained discriminative encoder to a restorative decoder (i.e., forming an encoder-decoder) for further combined discriminative and restorative learning (called Step (D)+R), and finally associating the pretrained autoencoder with an adversarial-encoder for the final full discriminative, restorative, and adversarial training (called Step ((D)+R)+A). This stepwise pretraining strategy provides the most reliable

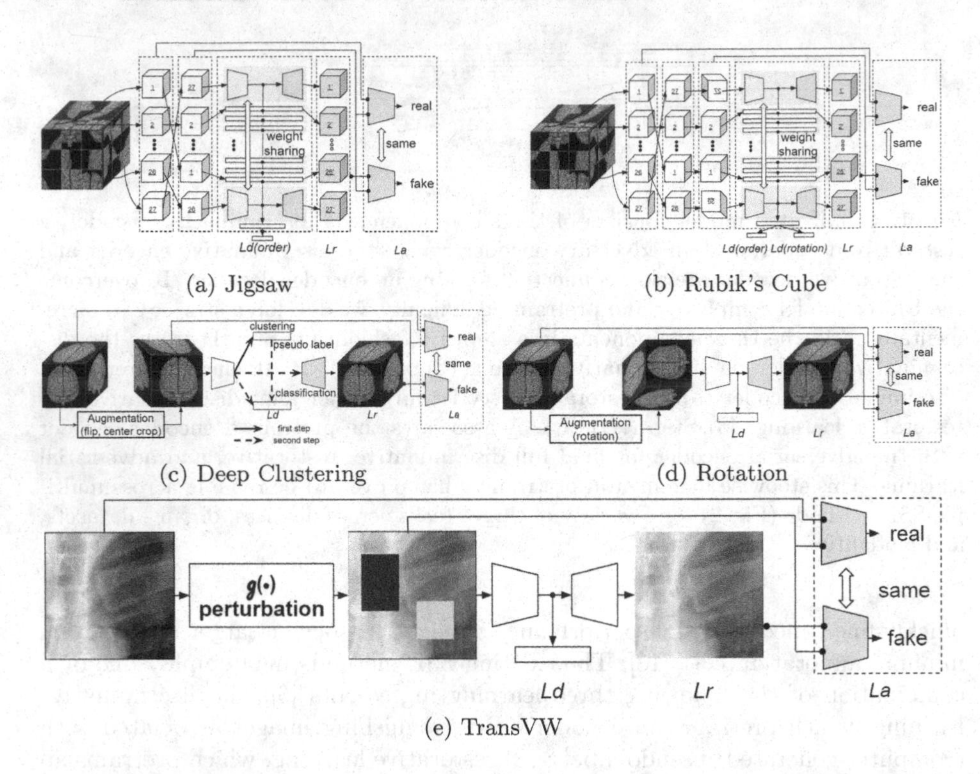

Fig. 2. Redesigning five prominent SSL methods: (a) Jigsaw, (b) Rubik's Cube, (c) Deep Clustering, (d) Rotation, and (e) TransVW in a United framework. The original Jigsaw [13], Deep Clustering [4], and Rotation [7] were proposed for 2D image analysis employing discriminative learning alone and provided only pretrained encoders; therefore, in our United framework (a, c, d), these methods have been augmented with two new components (in light blue) for restorative learning and adversarial learning and re-implemented in 3D. The code for the original Rubik's Cube [21] is not released and thus reimplemented and augmented with new learning ingredients in light blue (b). The original TransVW [8] is supplemented with adversarial learning (c). Following our redesign, all five methods provide all three learned components: discriminative encoders, restorative decoders, and adversary encoders, which are transferable to target classification and segmentation tasks. (Color figure online)

performance across most target tasks evaluated in this work encompassing both classification and segmentation (see Table 2 and 3 as well as Table 4 in the Supplementary Material).

Through our extensive experiments, we have observed that (1) discriminative learning alone (i.e., Step D) significantly enhances discriminative encoders on target classification tasks (e.g., +3% and +4% AUC improvement for lung nodule and pulmonary embolism false positive reduction as shown in Table 2) relative to training from scratch; (2) in comparison with (sole) discriminative learning, incremental restorative pretraining combined with continual discriminative learning (i.e., Step (D)+R) enhances discriminative encoders further for target classification tasks (e.g., +2% and +4% AUC improvement for lung nodule and pulmonary embolism false positive reduction as shown in Table 2) and boosts encoder-decoder models for target segmentation tasks (e.g., +3%, +7%, and +5% IoU improvement for lung nodule, liver, and brain tumor segmentation as shown in Table 3); and (3) compared with Step (D)+R, the final stepwise incremental pretraining (i.e., Step: ((D)+R)+A) generates sharper and more realistic medical images (e.g., FID decreases from 427.6 to 251.3 as shown in Table 5 in the Supplementary Material) and further strengthens each component for representation learning, leading to considerable performance gains (see Fig. 3) and annotation cost reduction (e.g., 28%, 43%, and 26% faster for lung nodule false positive reduction, lung nodule tumor segmentation, and pulmonary embolism false positive reduction as shown in Fig. 4) for five target tasks across diseases, organs, datasets, and modalities.

We should note that recently Haghighi et al. [9] also combined discriminative, restorative, and adversarial learning, but our findings complement theirs, and more importantly, our method significantly differs from theirs, because they were more concerned with contrastive learning (e.g., MoCo-v2 [5], Barlow Twins [19], and SimSiam [6]) and focused on 2D medical image analysis. By contrast, we are focusing on 3D medical imaging by redesigning five popular SSL methods beyond contrastive learning. As they acknowledged [9], their results on TransVW [8] augmented with an adversarial encoder were based on the experiments presented in this paper. Furthermore, this paper focuses on a stepwise incremental pretraining to stabilize United model training, revealing new insights into synergistic effects and contributions among the three learning ingredients.

In summary, we make the following three main contributions:

1. A stepwise incremental pretraining strategy that stabilizes United models' pretraining and unleashes the synergistic effects of the three SSL ingredients;
2. A collection of pretrained United models that integrate discriminative, restorative, and adversarial learning in a single framework for 3D medical imaging, encompassing both classification and segmentation tasks;
3. A set of extensive experiments that demonstrate how various pretraining strategies benefit target tasks across diseases, organs, datasets, and modalities.

2 Stepwise Incremental Pretraining

We have redesigned five prominent SSL methods, including Rotation, Jigsaw, Rubik's Cube, Deep Clustering, and TransVW, and augmented each with the missing components under our United framework (Fig. 2). A United model (Fig. 1) is a skip-connected encoder-decoder associated with an adversary encoder. With our redesign, for the first time, all five methods have all three SSL components. We *incrementally* train United models component by component in a *stepwise* manner, yielding three learned transferable components: discriminative encoders, restorative decoders, and adversarial encoders. The pretrained discriminative encoder can be fine-tuned for target classification tasks; the pretrained discriminative encoder and restorative decoder, forming a skip-connected encoder-decoder network (i.e., U-Net [14,16]), can be fine-tuned for target segmentation tasks.

Discriminative learning trains a discriminative encoder D_θ, where θ represents the model parameters, to predict target label $y \in Y$ from input $x \in X$ by minimizing a loss function for $\forall x \in X$ defined as

$$\mathcal{L}_d = -\sum_{n=1}^{N}\sum_{k=1}^{K} y_{nk} \ln(p_{nk}) \tag{1}$$

where N is the number of samples, K is the number of classes, and p_{nk} is the probability predicted by D_θ for x_n belonging to Class k; that is, $p_n = D_\theta(x_n)$ is the probability distribution predicted by D_θ for x_n over all classes. In SSL, the labels are automatically obtained based on the properties of the input data, involving no manual annotation. All five SSL methods in this work have a discriminative component formulated as a classification task, while other discriminative losses can be used, such as contrastive losses in MoCo-v2 [5], Barlow Twins [19], and SimSiam [6].

Restorative learning trains an encoder-decoder $(D_\theta, R_{\theta'})$ to reconstruct an original image x from its distorted version $\mathcal{T}(x)$, where $\mathcal{T}$ is a distortion function, by minimizing pixel-level reconstruction error:

$$\mathcal{L}_r = \mathbb{E}_x\, L_2(x, R_{\theta'}(D_\theta(\mathcal{T}(x)))) \tag{2}$$

where $L_2(u, v)$ is the sum of squared pixel-by-pixel differences between u and v.

Adversarial learning trains an additional adversary encoder, $A_{\theta''}$, to help the encoder-decoder $(D_\theta, R_{\theta'})$ reconstruct more realistic medical images and in turn strengthen representation learning. The adversary encoder learns to distinguish fake image pair $(R_{\theta'}(D_\theta(\mathcal{T}(x))), \mathcal{T}(x))$ from real pair $(x, \mathcal{T}(x))$ via an adversarial loss:

$$\mathcal{L}_a = E_{x,\mathcal{T}(x)} log A_{\theta''}(\mathcal{T}(x), x) + E_x log(1 - A_{\theta''}(\mathcal{T}(x), R_{\theta'}(D_\theta(\mathcal{T}(x))))) \tag{3}$$

The final objective combines all losses:

$$\mathcal{L} = \lambda_d \mathcal{L}_d + \lambda_r \mathcal{L}_r + \lambda_a \mathcal{L}_a \tag{4}$$

where λ_d, λ_r, and λ_a controls the importance of each learning ingredients.

Stepwise incremental pretraining trains our United models continually component-by-component because training a whole United model in an end-to-end fashion (i.e., all three components together directly from scratch)—a strategy called (D+R+A)—is unstable. For example, as shown in Table 1, Strategy ((D)+R)+A) (see Fig. 1) always outperforms Strategy (D+R+A) and provides the most reliable performance across most target tasks evaluated in this work.

Table 1. ((D)+R)+A strategy always outperforms D+R+A strategy on all five target tasks. We report the mean and standard deviation across ten runs and performed independent two sample t-test between the two strategies. The text is bolded when they are significantly different at p = 0.05 level

Method	Approach	BMS	NCC	LCS	ECC	NCS
Jigsaw	(D+R+A)	64.98 ± 0.68	97.24 ± 0.73	83.54 ± 0.95	84.12 ± 1.38	74.32 ± 1.54
	(((D)+R)+A)	**66.07 ± 1.33**	97.86 ± 1.54	**84.87 ± 1.67**	84.89 ± 1.05	74.87 ± 1.17
Rubik's Cube	(D+R+A)	65.13 ± 1.34	98.21 ± 0.88	84.12 ± 1.19	84.36 ± 1.17	75.18 ± 1.32
	(((D)+R)+A)	**66.88 ± 1.72**	**99.17 ± 0.79**	**85.18 ± 0.99**	**85.64 ± 0.87**	76.07 ± 1.23
TransVW Adv.	(D+R+A)	66.81 ± 1.06	97.63 ± 0.52	85.16 ± 0.67	85.84 ± 1.84	76.32 ± 1.25
	(((D)+R)+A)	**69.57 ± 1.13**	**98.87 ± 0.61**	**86.85 ± 0.81**	**86.91 ± 3.27**	**77.51 ± 1.36**

3 Experiments and Results

Datasets and Metrics. To pretrain all five United models, we used 623 CT scans from the LUNA16 [15] dataset. We adopted the same strategy as [20], and cropped sub-volumes with a pixel size of 64 × 64 × 64. To evaluate the effectiveness of pretraining the five methods, we tested their performance on five 3D medical imaging tasks (See §B) including BraTS [2,12], LUNA16 [15], LiTS [3], PE-CAD [17], and LIDC-IDRI [1]. The acronyms BMS, LCS, and NCS denote the tasks of segmenting a brain tumor, liver, and lung nodules; NCC and

Table 2. Discriminative learning alone or combined with incremental restorative learning enhance discriminative encoders for classification tasks. We report the mean and standard deviation (mean ± s.d.) across ten trials, along with the statistic analysis (*p <0.5, **p <0.1, ***p <0.05) with and without incremental restorative pretraining for five self-supervised learning methods. In the "Decoder" column, ✗ and ✓ denote a non-pretrained decoder and with pretrained decoders but not used for target tasks. With incremental restorative learning, the performance gains were consistent for both target tasks.

Method	Approach	Decoder	NCC	ECC
Random	-	✗	94.25 ± 5.07	79.99 ± 8.06
Jigsaw	D	✗	95.49 ± 1.24	81.79 ± 1.04
	(D)+R	✓	97.29 ± 1.09***	84.39 ± 1.47***
Rubik's cube	D	✗	96.24 ± 1.57	81.76 ± 1.32
	(D)+R	✓	98.14 ± 0.38***	84.14 ± 1.58***
Deep clustering	D	✗	97.27 ± 1.43	84.82 ± 0.62
	(D)+R	✓	98.11 ± 0.55	85.12 ± 1.37
TransVW	D	✗	97.49 ± 0.45	84.25 ± 3.91
	(D)+R	✓	98.47 ± 0.22*	87.07 ± 2.83*
Rotation	D	✗	96.13 ± 2.41	82.37 ± 1.64
	(D)+R	✓	97.17 ± 0.81	83.57 ± 1.21*

Table 3. Incremental restorative learning ((D)+R) directly boost target segmentation tasks. In the "Decoder" column, ✗, ✓, and ✓ denote a non-pretrained decoder, not using pretrained decoders, and using pretrained decoders, respectively. Statistic analysis (*p <0·5, **p <0·1, ***p <0·05) was conducted between ✗ and ✓.

Method	Approach	Decoder	NCS	LCS	BMS
Random	-	✗	74.05 ± 1.97	77.82 ± 3.87	58.52 ± 2.61
Jigsaw	D	✗	73.38 ± 1.65	82.04 ± 1.65	63.33 ± 1.11
	(D)+R	✓	73.58 ± 1.26	83.04 ± 1.21	64.17 ± 0.62
	(D)+R	✓	**74.53 ± 1.13***	**84.17 ± 1.48*** **	**65.33 ± 1.31*** **
Rubik's Cube	D	✗	72.87 ± 0.86	77.42 ± 0.43	62.75 ± 1.93
	(D)+R	✓	74.33 ± 1.83	84.21 ± 0.24	64.91 ± 0.76
	(D)+R	✓	**75.66 ± 0.74*** **	**85.02 ± 1.08*** **	**65.83 ± 1.16*** **
Deep Clustering	D	✗	74.82 ± 0.47	82.67 ± 0.69	65.81 ± 0.73
	(D)+R	✓	75.01 ± 0.69	83.75 ± 0.9	66.14 ± 0.87
	(D)+R	✓	**75.91 ± 1.12*** **	**84.63 ± 0.63*** **	**66.73 ± 0.51*** **
TransVW	D	✗	76.93 ± 0.87	85.09 ± 2.15	64.02 ± 0.98
	(D)+R	✓	77.09 ± 1.52	85.63 ± 0.96	67.52 ± 0.87
	(D)+R	✓	**77.33 ± 0.52**	**86.53 ± 1.31***	**68.82 ± 0.38*** **
Rotation	D	✗	74.24 ± 0.91	82.44 ± 1.45	63.98 ± 0.84
	(D)+R	✓	74.65 ± 1.26	83.24 ± 2.21	64.54 ± 1.36
	(D)+R	✓	**74.86 ± 0.58***	**84.65 ± 1.01*** **	**65.44 ± 0.67*** **

ECC denote the tasks of reducing lung nodule and pulmonary embolism false positives results, respectively. We measured the performances of the pretrained models on five target tasks and reported the AUC (Area Under the ROC Curve) for classification tasks and IoU (Intersection over Union) for segmentation tasks. All target tasks ran at least 10 times and statistical analysis was performed using independent two-sample t-test.

(1) Incremental restorative learning ((D)+R) enhances discriminative encoders further for classification tasks. After pretraining discriminative encoders, we append restorative decoders to the end of the encoders and continue to pretrain discriminative encoder and restorative decoder together. The incremental restorative learning significantly enhances encoders in classification tasks, as shown in Table 2. Specifically, compared with the original methods, the incremental restorative learning improves Jigsaw by AUC scores of 1.9% and 2.6% in NCC and ECC; similarly, it improves Rubik's Cube by 1.9% and 2.4%, Deep Clustering by 0.9% and 0.3%, TransVW by 1.0% and 2.9%, and Rotation by 1.0% and 1.2%. The discriminative encoders are enhanced because they not only learn global features for discriminative tasks but also learns fine-grained features through incremental restorative learning.

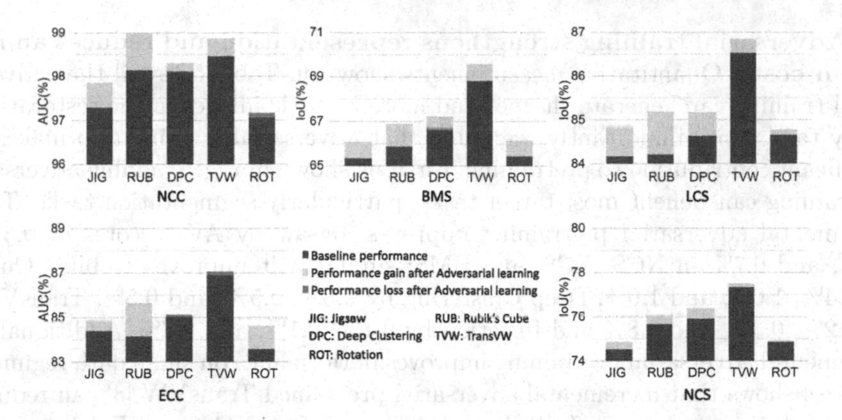

Fig. 3. Adversarial training strengthens learned representation. Target tasks performance are generally increased (red) following the adversarial training. Although some target tasks show a decrease (pink), these reductions do not reach statistical significance according to the t-test. (Color figure online)

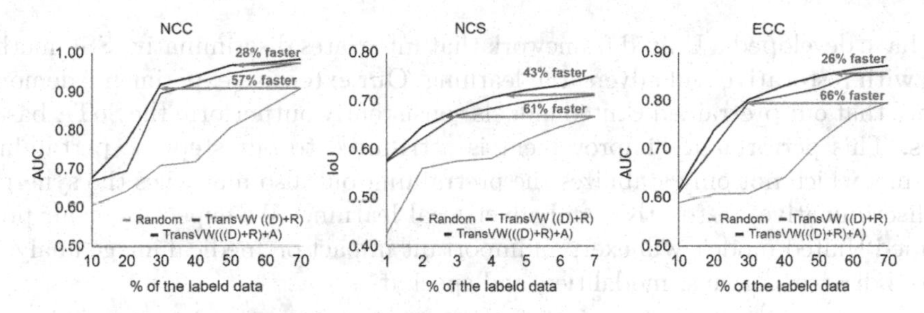

Fig. 4. Adversarial training reduces annotation costs. TransVW combined with adversarial learning reduces the annotation cost by 28%, 43%, and 26% for target tasks of NCC, NCS, and ECC, respectively, compared with the original TransVW. It also reduces the annotation cost by 57%, 61%, and 66% for target tasks of NCC, NCS, and ECC, respectively, compared with initializing the network from random.

(2) Incremental restorative learning ((D)+R) directly boost target segmentation tasks. Most state-of-the-art segmentation methods do not pretrain their decoders but instead initialize them at random [5,10]. We argue that the random decoders are suboptimal, evidenced by the data in Table 3, and we demonstrate that the incremental pretrained restorative decoders can directly boost target segmentation tasks. In particular, compared with the original methods, the incremental pretrained restorative decoder improves Jigsaw by 1.2%, 2.1% and 2.0% IoU improvement in NCS, LCS and BMS; similarly, it improves Rubik's Cube by 2.8%, 7.6%, and 3.1%, Deep Clustering by 1.1%, 2.0%, and 0.9%, TransVW by 0.4%, 1.4%, and 4.8% and Rotation by 0.6%, 2.2% and 1.5%. The consistent performance gain suggests that a wide variety of target segmentation tasks can benefit from our incremental pretrained restorative decoders.

(3) Adversarial training strengthens representation and reduces annotation costs. Quantitative measurements shown in Table 5 reveal that adversarial training can generate sharper and more realistic images in the restoration proxy task. More importantly, we found that adversarial training also makes a significant contribution to pretraining. First, as shown in Fig. 3, adding adversarial training can benefit most target tasks, particularly segmentation tasks. The incremental adversarial pretraining improves Jigsaw by AUC scores of 0.3%, 0.7%, and 0.7% in NCS, LCS, and BMS; similarly, it improves Rubik's Cube by 0.4%, 1.0%, and 1.0%, Deep Clustering by 0.5%, 0.5%, and 0.5%, TransVW by 0.2%, 0.3%, and 0.8% and Rotation by 0.1%, 0.1%, and 0.7%. Additionally, incremental adversarial pretraining improves performance on small data regimes. Figure 4 shows that incremental adversarial pretrained TransVW [8] can reduce the annotation cost by 28%, 43%, and 26% on NCC, NCS, and ECC, respectively, compared with TransVW [8].

4 Conclusion

We have developed a United framework that integrates discriminative SSL methods with restorative and adversarial learning. Our extensive experiments demonstrate that our pretrained United models consistently outperform the SoTA baselines. This performance improvement is attributed to our stepwise pertaining scheme, which not only stabilizes the pretraining but also unleashes the synergy of discriminative, restorative, and adversarial learning. We expect that our pretrained United models will exert an important impact on medical image analysis across diseases, organs, modalities, and specialties.

Acknowledgments. We thank F. Haghighi, M. R. Hosseinzadeh Taher, and Z. Zhou for their discussions, debates, and supports in implementing the earlier ideas behind "United & Unified" and in drafting earlier versions. This research has been supported in part by ASU and Mayo Clinic through a Seed Grant and an Innovation Grant, and in part by the NIH under Award Number R01HL128785. The content is solely the responsibility of the authors and does not necessarily represent the official views of the NIH. This work has utilized the GPUs provided in part by the ASU Research Computing and in part by the Extreme Science and Engineering Discovery Environment (XSEDE) funded by the National Science Foundation (NSF) under grant numbers: ACI-1548562, ACI-1928147, and ACI-2005632. The content of this paper is covered by patents pending.

References

1. Armato III, S.G., et al.: The lung image database consortium (LIDC) and image database resource initiative (IDRI): a completed reference database of lung nodules on CT scans. Med. Phys. **38**(2), 915–931 (2011)

2. Bakas, S., et al.: Identifying the best machine learning algorithms for brain tumor segmentation, progression assessment, and overall survival prediction in the brats challenge. arXiv preprint arXiv:1811.02629 (2018)
3. Bilic, P., et al.: The liver tumor segmentation benchmark (LiTS). arXiv preprint arXiv:1901.04056 (2019)
4. Caron, M., Bojanowski, P., Joulin, A., Douze, M.: Deep clustering for unsupervised learning of visual features. In: European Conference on Computer Vision (2018)
5. Chen, X., Fan, H., Girshick, R., He, K.: Improved baselines with momentum contrastive learning (2020)
6. Chen, X., He, K.: Exploring simple Siamese representation learning. In: Proceedings of the IEEE/CVF Conference on Computer Vision and Pattern Recognition (CVPR), pp. 15750–15758, June 2021
7. Gidaris, S., Singh, P., Komodakis, N.: Unsupervised representation learning by predicting image rotations (2018)
8. Haghighi, F., Taher, M.R.H., Zhou, Z., Gotway, M.B., Liang, J.: Transferable visual words: exploiting the semantics of anatomical patterns for self-supervised learning. IEEE Trans. Med. Imaging, 1 (2021). https://doi.org/10.1109/TMI.2021.3060634
9. Haghighi, F., Taher, M.R.H., Gotway, M.B., Liang, J.: DiRA: discriminative, restorative, and adversarial learning for self-supervised medical image analysis. In: Proceedings of the IEEE/CVF Conference on Computer Vision and Pattern Recognition, pp. 20824–20834 (2022)
10. He, K., Fan, H., Wu, Y., Xie, S., Girshick, R.: Momentum contrast for unsupervised visual representation learning (2020)
11. Jing, L., Tian, Y.: Self-supervised visual feature learning with deep neural networks: a survey. IEEE Trans. Pattern Anal. Mach. Intell. **43**(11), 4037–4058 (2020)
12. Menze, B.H., et al.: The multimodal brain tumor image segmentation benchmark (BRATS). IEEE Trans. Med. Imaging **34**(10), 1993–2024 (2014)
13. Noroozi, M., Favaro, P.: Unsupervised learning of visual representations by solving jigsaw puzzles. In: Leibe, B., Matas, J., Sebe, N., Welling, M. (eds.) ECCV 2016. LNCS, vol. 9910, pp. 69–84. Springer, Cham (2016). https://doi.org/10.1007/978-3-319-46466-4_5
14. Ronneberger, O., Fischer, P., Brox, T.: U-Net: convolutional networks for biomedical image segmentation. In: Navab, N., Hornegger, J., Wells, W.M., Frangi, A.F. (eds.) MICCAI 2015. LNCS, vol. 9351, pp. 234–241. Springer, Cham (2015). https://doi.org/10.1007/978-3-319-24574-4_28
15. Setio, A.A.A., et al.: Validation, comparison, and combination of algorithms for automatic detection of pulmonary nodules in computed tomography images: the LUNA16 challenge. Med. Image Anal. **42**, 1–13 (2017)
16. Siddique, N., Sidike, P., Elkin, C., Devabhaktuni, V.: U-Net and its variants for medical image segmentation: theory and applications (2020). http://arxiv.org/abs/2011.01118
17. Tajbakhsh, N., Gotway, M.B., Liang, J.: Computer-aided pulmonary embolism detection using a novel vessel-aligned multi-planar image representation and convolutional neural networks. In: Navab, N., Hornegger, J., Wells, W.M., Frangi, A.F. (eds.) MICCAI 2015. LNCS, vol. 9350, pp. 62–69. Springer, Cham (2015). https://doi.org/10.1007/978-3-319-24571-3_8
18. Tajbakhsh, N., Roth, H., Terzopoulos, D., Liang, J.: Guest editorial annotation-efficient deep learning: the holy grail of medical imaging. IEEE Trans. Med. Imaging **40**(10), 2526–2533 (2021)
19. Zbontar, J., Jing, L., Misra, I., LeCun, Y., Deny, S.: Barlow twins: self-supervised learning via redundancy reduction. arXiv:2103.03230 (2021)

20. Zhou, Z., Sodha, V., Pang, J., Gotway, M.B., Liang, J.: Models genesis. Med. Image Anal. **67**, 101840 (2021). https://doi.org/10.1016/j.media.2020.101840
21. Zhuang, X., Li, Y., Hu, Y., Ma, K., Yang, Y., Zheng, Y.: Self-supervised feature learning for 3D medical images by playing a Rubik's cube. In: Shen, D., et al. (eds.) MICCAI 2019. LNCS, vol. 11767, pp. 420–428. Springer, Cham (2019). https://doi.org/10.1007/978-3-030-32251-9_46

POPAR: Patch Order Prediction and Appearance Recovery for Self-supervised Medical Image Analysis

Jiaxuan Pang[1], Fatemeh Haghighi[1], DongAo Ma[1], Nahid Ul Islam[1], Mohammad Reza Hosseinzadeh Taher[1], Michael B. Gotway[2], and Jianming Liang[1](✉)

[1] Arizona State University, Tempe, AZ 85281, USA
{jpang12,fhaghigh,dongaoma,nuislam,mhossei2,jianming.liang}@asu.edu
[2] Mayo Clinic, Scottsdale, AZ 85259, USA
Gotway.Michael@mayo.edu

Abstract. Vision transformer-based self-supervised learning (SSL) approaches have recently shown substantial success in learning visual representations from unannotated photographic images. However, their acceptance in medical imaging is still lukewarm, due to the significant discrepancy between medical and photographic images. Consequently, we propose POPAR (patch order prediction and appearance recovery), a novel vision transformer-based self-supervised learning framework for chest X-ray images. POPAR leverages the benefits of vision transformers and unique properties of medical imaging, aiming to simultaneously learn patch-wise high-level contextual features by correcting shuffled patch orders and fine-grained features by recovering patch appearance. We transfer POPAR pretrained models to diverse downstream tasks. The experiment results suggest that (1) POPAR outperforms state-of-the-art (SoTA) self-supervised models with vision transformer backbone; (2) POPAR achieves significantly better performance over all three SoTA contrastive learning methods; and (3) POPAR also outperforms fully-supervised pretrained models across architectures. In addition, our ablation study suggests that to achieve better performance on medical imaging tasks, both fine-grained and global contextual features are preferred. All code and models are available at GitHub.com/JLiangLab/POPAR.

Keywords: Vision transformer · Self-supervised learning · Medical image analysis · Transfer learning

1 Introduction

Self-supervised learning (SSL) aims to learn generalizable representations from (unannotated) images and transfer the learned representations to application-

Supplementary Information The online version contains supplementary material available at https://doi.org/10.1007/978-3-031-16852-9_8.

K. Kamnitsas et al. (Eds.): DART 2022, LNCS 13542, pp. 77–87, 2022.
https://doi.org/10.1007/978-3-031-16852-9_8

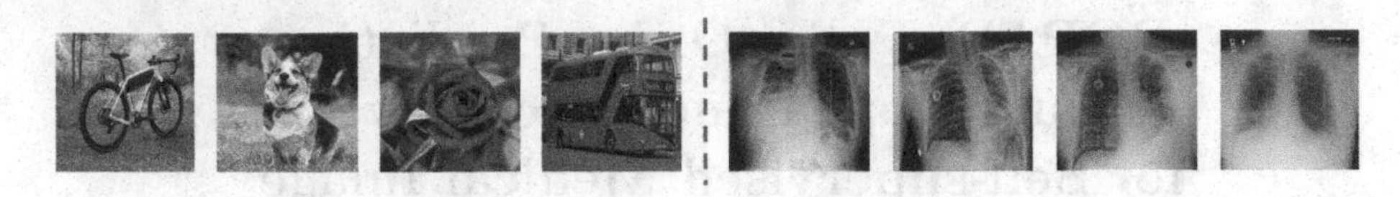

Fig. 1. Photographic images typically have objects of considerable differences (bicycle, dog, flower, etc.) centered in front of varying backgrounds, while medical images generated from a particular imaging protocol are remarkably similar in anatomy (e.g., lung) across patients with diagnostic information spread across entire images (e.g., conditions as boxed in yellow). Analyzing medical images requires not only high-level knowledge of anatomical structures and their relationships but also fine-grained features across entire images. Our POPAR aims to meet this requirement by autodidactically learning high-level anatomical knowledge via patch order prediction and automatically gleaning fine-grained features via (patch) appearance recovery (see Fig. 2). (Color figure online)

specific tasks to boost performance and reduce annotation efforts [19]. SSL has achieved state-of-the-art (SoTA) performance, sometimes even surpassing standard supervised ImageNet models in computer vision [5,6,26]. However, its popularity in medical imaging remains tepid, even in light of annotation dearth, a significant challenge facing deep learning for medical image analysis (MIA) [23]. We believe that this is due to the marked differences between medical and photographic images [12,22,27]. Photographic images, particularly those in ImageNet [8], typically have objects of considerable variations (cats, dogs, flowers, etc.), with distinctive components, centered in front of varying backgrounds (see Fig. 1). Therefore, object recognition in photographic images is based mainly on high-level features extracted from objects' discriminative components [12,22]. By contrast, medical imaging protocols are designed for specified clinical purposes by focusing on particular body parts, generating images of remarkable similarity in anatomy across patients [14]. For example, the posteroanterior chest X-rays all look similar (Fig. 1). However, diagnostically valuable information may spread across entire images. Therefore, understanding high-level anatomical structures and their relative spatial relationships is essential for distinguishing diseases from normal anatomy [12]. Nevertheless, the fine-grained details throughout entire images are equally indispensable because identifying diseases, delineating organs, and isolating lesions may rely on subtle texture variations [12,16]. Therefore, a natural question is: *how to learn integrated high-level and fine-grained features from medical images via self-supervision?*

To answer this question, we have developed a new SSL method called POPAR (Patch Order Prediction and Appearance Recovery), because it is equipped with two novel learning perspectives: (1) patch order prediction, which autodidactically learns high-level anatomical structures and their relative relationships, and (2) (patch) appearance recovery, which automatically gleans fine-grained features from medical images. We employ Swin Transformer as the POPAR's backbone because its hierarchical design enables multi-scale modeling, which naturally supports the two learning perspectives simultaneously.

For performance comparison and ablation studies, we have also trained three downgraded versions of POPAR: POPAR^{-1}, POPAR^{-2}, and POPAR^{-3} (see Table 1). Our extensive experiments demonstrate that (1) POPAR outperforms self-supervised ImageNet models with transformer backbone (see Table 2); (2) POPAR outperforms SoTA self-supervised pretrained models with CNN and transformer backbones (see Table 3); and (3) POPAR outperforms fully-supervised pretrained models across CNN and transformer architectures (see Table 4). This performance is attributed to our insights into the requirements of medical imaging tasks for global anatomical knowledge and fine-grained details in texture variations (see Sect. 5: Pretraining tasks).

In summary, we make the following main contributions:

1. A novel vision transformer-based SSL framework that simultaneously learns global relationships of anatomical structures and fine-grained details embedded in medical images.
2. A collection of pretrained models for transformer architectures (ViT-B and Swin-B) that yield SoTA performance on a set of MIA classification tasks.
3. An extensive set of experiments that demonstrate POPAR's superiority over SoTA supervised and self-supervised pretrained models across architectures.

2 Related Works and Novelties

Image Context Learning. Image context has been shown to be a powerful source for learning visual representations via SSL. Multiple pretext tasks have been formulated to predict the context arrangement of image patches, including predicting the relative position of two image patches [10], solving Jigsaw puzzles [20], and playing Rubik's cube [28]. These methods employ multi-Siamese CNN backbones as feature extractors, followed by additional feature aggregation layers for determining the relationships between the input patches. However, the feature aggregation layers are discarded after the pretraining step, and only the pretrained multi-Siamese CNNs are transferred to the target tasks. As a result, the learned relationships among image patches are mainly ignored in the target tasks. In contrast to these approaches, our POPAR uses the multi-head attention mechanism to capture the relationships among anatomical patterns embedded in image patches, which is fully transferable to target tasks.

Masked Image Modeling. Inspired by masked language modeling [3,9], multiple vision transformer-based SSL methods have been developed for masked image modeling. BEiT [2] predicts the discrete tokens from masked images. SimMIM [25] and MAE [15] mask random patches from the input image and reconstruct the missing patches. While POPAR bears similarities to these methods in patch reconstruction, it distinguishes itself from them by (1) reconstructing correct image patches from misplaced patches or from transformed patches, and (2) predicting the correct positions of shuffled image patches for learning global contextual features.

Restorative Learning. The restorative SSL methods aim to learn representations by recovering original images from their distorted versions. Multiple SSL methods have incorporated image restoration into their pretext tasks. Models Genesis [27] proposed four effective image transformations for restorative SSL in medical imaging. TransVW [13,14] introduced a SSL framework for learning semantic representation from the consistent anatomical structures. CAiD [22] formulated a restoration task to boost instance discrimination SSL with context-aware representations. DiRA [12] integrates discriminative, restorative, and adversarial SSL to learn fine-grained representations via collaborative learning. However, none of these approaches learns anatomical relationships among image patches. By contrast, POPAR employs a transformer backbone to integrate restorative learning with patch order prediction, capturing not only visual details but also relationships among anatomical structures.

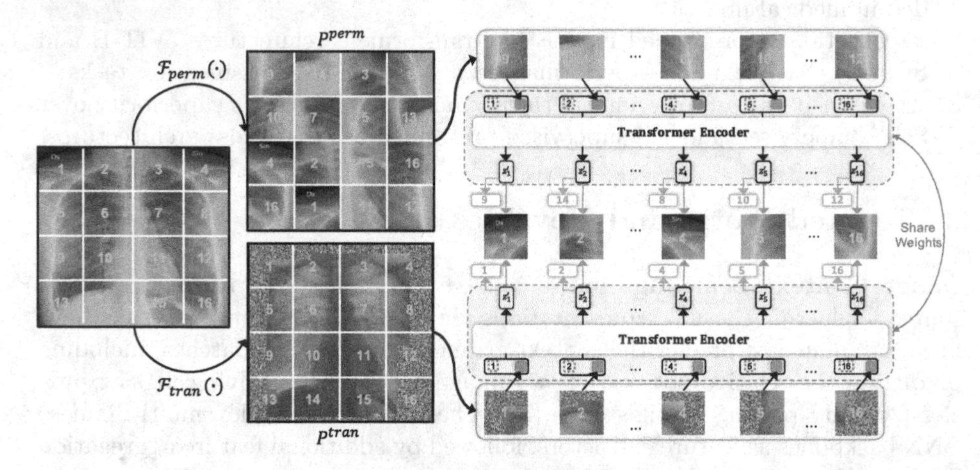

Fig. 2. POPAR aims to learn (1) contextualized high-level anatomical structures via patch order prediction, and (2) fine-grained image features via patch appearance recovery. For each image, we divide it into a sequence of non-overlapping patches, and randomly distort the patch order (upper path) or patch appearances (bottom path). We give the distorted patch sequence to a transformer network, and train the model to predict the correct position of each input patch and recover the correct patch appearance for each position as the original patch sequence.

3 Method

Notations. Given an image sample $x \in \mathbb{R}^{H \times W \times C}$, where (H, W) is the resolution of the image and C is the number of channels, we randomly select and apply one of the following distortion functions: (a) patch order distortion $\mathcal{F}_{perm}(\cdot)$ (the upper path in Fig. 2) or (b) patch appearance distortion $\mathcal{F}_{tran}(\cdot)$ (the bottom path in Fig. 2). In patch order distortion, we first divide x into a sequence of n non-overlapping image patches $P = (p_1, p_2, ..., p_n)$, where $n = \frac{H \times W}{k^2}$ and (k, k) is the resolution of each patch. We use $L = (1, 2, ..., n)$

to denote the correct patch positions within x. We then apply a random permutation operator on L to generate permuted patch positrons L^{perm}. We use L^{perm} to re-arrange the patch sequence P, resulting in permuted patch sequence P^{perm}. In patch appearance distortion, we first apply an image transformation operator on x, resulting in an appearance-transformed image x^{tran}. We then divide x^{tran} into a sequence of n non-overlapping *transformed* image patches $P^{tran} = (p_1^{tran}, p_2^{tran}, ..., p_n^{tran})$. Following [11], we map the patches in P^{perm} and P^{tran} to D dimension patch embeddings using a trainable linear projection layer. Then, trainable positional embeddings are added to the patch embeddings, resulting in a sequence of embedding vectors. The embedding vectors are further processed by the transformer encoder $g_\theta(\cdot)$ to generate a set of contextual patch features $Z' = (z'_1, z'_2, ..., z'_n)$. We then pass Z' to two distinct prediction heads $s_\theta(\cdot)$ and $k_\theta(\cdot)$ to generate predictions $\mathcal{P}^{pop} = s_\theta(Z')$ and $\mathcal{P}^{ar} = k_\theta(Z')$ for performing the patch order prediction and patch appearance recovery, respectively, as described below. Following [21], we define $\stackrel{!}{=}$ to be "shall be (made) equal".

Patch order prediction aims to predict the correct position of a patch based on its appearance. Particularly, depending on which distortion function is selected, the expected prediction for $\mathcal{P}^{pop}$ is formulated as follows.

$$\begin{cases} \mathcal{P}^{pop} \stackrel{!}{=} L_{perm} & \text{if } \mathcal{F}_{perm}(\cdot) \text{ is selected} \\ \mathcal{P}^{pop} \stackrel{!}{=} L & \text{if } \mathcal{F}_{tran}(\cdot) \text{ is selected} \end{cases} \tag{1}$$

Patch appearance recovery aims to reconstruct the correct appearance for each position in the input sequence. We expect the network to predict the original appearance in P regardless of which distortion function ($\mathcal{F}_{perm}(\cdot)$ or $\mathcal{F}_{tran}(\cdot)$) is selected. The expected reconstruction prediction for $\mathcal{P}^{ar}$ is defined as follows.

$$\mathcal{P}^{ar} \stackrel{!}{=} P \tag{2}$$

Overall Training Scheme. We formulate the patch order prediction as a n-way multi-class classification task and optimize the model by minimizing the categorical cross-entropy loss: $\mathcal{L}_{pop} = -\frac{1}{B}\sum_{b=1}^{B}\sum_{l=1}^{n}\sum_{c=1}^{n} \mathcal{Y}_{blc} log \mathcal{P}_{blc}^{pop}$, where B denotes the batch size, n is the number of patches for each image, $\mathcal{Y}$ represent the ground truth (as defined in Eq. (1)), and $\mathcal{P}^{pop}$ represents the network's patch order prediction. Moreover, we formulate the patch appearance recovery as a reconstruction task and train the model by minimizing $L2$ distance between the original patch sequence P and the restored patch sequence $\mathcal{P}^{ar}$: $\mathcal{L}_{ar} = \frac{1}{B}\sum_{i=1}^{B}\sum_{j=1}^{n} ||p_j - p_j^{ar}||_2^2$, where p_j and p_j^{ar} represent the patch appearance from P and $\mathcal{P}^{ar}$, respectively. We integrate both learning schemes and train POCAR with an overall loss function $\mathcal{L}_{popar} = \lambda * \mathcal{L}_{pop} + (1 - \lambda) * \mathcal{L}_{ar}$, where λ is the weight to specify the importance of each loss. The formulation of the $\mathcal{L}_{pop}$ encourages the transformer model to learn high-level anatomical structures and their relative relationships. Moreover, the definition of $\mathcal{L}_{ar}$ encourages the model to capture more fine-grained features from images.

4 Experiments and Results

4.1 Implementation Details

Pretraining Settings. We pretrain POPAR with ViT-B and Swin-B as backbones using their official default configurations on the training set of ChestX-ray14 [24] dataset. Due to architecture differences (detailed in appendix), we use image size of 224×224 and 448×448 for ViT-B and Swin-B backbones, respectively. Accordingly, we divide images into 16×16 and 32×32 patches for ViT-B and Swin-B, respectively, which results in $n = 196$ patches in both backbones. We use two single linear layers as the prediction heads for the classification (order prediction) and restoration (appearance recovery) tasks. For all models, we use SGD optimizer with learning rate 0.1. We set λ to 0.5. We train POPAR models with ViT-B and Swin-B backbones for 1000 and 300 epochs, respectively. Image transformation function $\mathcal{F}_{tran}(\cdot)$ includes local pixel shuffling, non-linear transformation, and outer/inner cutouts [27]. More details are in the appendix.

Table 1. We evaluate POPAR with Swin-B and ViT-B backbones using four different pretraining and finetuning image resolutions, denoted as PT and FT, respectively. POPAR, our official implementation, is the model with Swin-B backbone, pretraining and finetuning resolutions of 448×448, which yields the best performance on all target tasks. For performance comparison and ablation studies, we have pretrained three downgraded versions: (1) POPAR^{-1} with Swin-B backbone, pretraining resolution of 448×448, and finetuning resolution of 224×224; (2) POPAR^{-2} with Swin-B backbone and pretraining and finetuning resolutions of 224×224; and (3) POPAR^{-3} with ViT-B backbone and pretraining and finetuning resolutions of 224×224.

Setup name	Backbone	Shuffled patches	PT → FT	ChestX-ray14	CheXpert	ShenZhen	RSNA Pneumonia
POPAR^{-3}	ViT-B	196	$224^2 \rightarrow 224^2$	79.58 ± 0.13	87.86 ± 0.17	93.87 ± 0.63	73.17 ± 0.46
POPAR^{-2}	Swin-B	47	$224^2 \rightarrow 224^2$	79.50 ± 0.20	87.63 ± 0.39	95.07 ± 1.22	73.07 ± 0.46
POPAR^{-1}		196	$448^2 \rightarrow 224^2$	80.51 ± 0.15	88.25 ± 0.78	96.81 ± 0.40	73.58 ± 0.18
POPAR		196	$448^2 \rightarrow 448^2$	$\mathbf{81.81 \pm 0.10}$	$\mathbf{88.34 \pm 0.50}$	$\mathbf{97.33 \pm 0.74}$	$\mathbf{74.19 \pm 0.37}$

Target Tasks and Finetuning Settings. We evaluate the efficacy of POPAR models in transfer learning to four medical classification tasks in chest X-ray datasets, including ChestX-ray14, CheXpert [17], NIH Shenzhen CXR [18], and RSNA Pneumonia [1,24]. We transfer POPAR models to target tasks by removing the prediction heads and inserting randomly initialized target classification heads that include (1) a linear layer for the ViT-B backbone and (2) an average pooling and a linear layer for the Swin-B backbone. We finetune all the parameters of target models. Details of target tasks and finetuning settings are provided in the appendix.

4.2 Results

(1) POPAR outperforms self-supervised ImageNet models with transformer backbone. To demonstrate the effectiveness of pretraining transformers with in-domain medical data, we compare POPAR with SoTA transformer-based self-supervised methods that are pretrained on ImageNet. We evaluate existing self-supervised ImageNet models with ViT-B (MoCoV3 [7], SimMIM [25], DINO [4], BEiT [2], and MAE [15]) and Swin-B (SimMIM) backbones. We use officially released models for all baselines, among which BEiT model is pretrained on the ImageNet-21K dataset, while the rest of the models are pretrained on the ImageNet-1K dataset. We make the following observations from the results in Table 2. Firstly, SimMIM and MAE achieve superior performance over other baselines, demonstrating the effectiveness of masked image restoration for pretraining transformer models. Secondly, POPAR with ViT-B backbone surpasses all self-supervised ImageNet models with the same backbone. Thirdly, even POPAR^{-1} (a downgraded version of POPAR) outperforms SimMIM with Swin-B backbone on three out of four target tasks.

Table 2. Even POPAR^{-1} and POPAR^{-3} (two downgraded versions of POPAR) outperform SoTA self-supervised ImageNet models with transformer backbone in three target tasks. The best methods are bolded, while the second best are underlined.

Backbone	Method	ChestX-ray14	CheXpert	ShenZhen	RSNA Pneumonia
ViT-B	MoCoV3	79.20 ± 0.29	86.91 ± 0.77	85.71 ± 1.41	72.79 ± 0.52
	SimMIM	79.55 ± 0.56	87.83 ± 0.46	92.74 ± 0.92	72.08 ± 0.47
	DINO	78.37 ± 0.47	86.91 ± 0.44	87.83 ± 7.20	71.27 ± 0.45
	BEiT	74.69 ± 0.39	85.81 ± 1.00	92.95 ± 1.25	72.78 ± 0.37
	MAE	78.97 ± 0.65	87.12 ± 0.54	93.58 ± 1.18	72.85 ± 0.50
	POPAR^{-3}	79.58 ± 0.13	<u>87.86 ± 0.17</u>	<u>93.87 ± 0.63</u>	<u>73.17 ± 0.46</u>
Swin-B	SimMIM	**81.39 ± 0.18**	87.50 ± 0.23	87.86 ± 4.92	73.15 ± 0.73
	POPAR^{-1}	<u>80.51 ± 0.15</u>	**88.16 ± 0.66**	**96.81 ± 0.40**	**73.58 ± 0.18**

Table 3. Even POPAR^{-1} (a downgraded version of POPAR) yields significant performance boosts ($p < 0.05$) in comparison with SoTA self-supervised methods pretrained on ResNet-50 or transformer architectures. All models are pretrained on the ChestX-ray14 dataset. The best methods are bolded while the second best are underlined.

Backbone	Method	ChestX-ray14	CheXpert	ShenZhen	RSNA Pneumonia
ResNet-50	SimSiam	79.62 ± 0.34	83.82 ± 0.94	93.13 ± 1.36	71.20 ± 0.60
	MoCoV2	80.36 ± 0.26	86.42 ± 0.42	92.59 ± 1.79	71.98 ± 0.82
	Barlow Twins	<u>80.45 ± 0.29</u>	86.90 ± 0.62	92.17 ± 1.54	71.45 ± 0.82
ViT-B	SimMIM	79.20 ± 0.19	83.48 ± 2.43	93.77 ± 1.01	71.66 ± 0.75
	POPAR^{-3}	79.58 ± 0.13	<u>87.86 ± 0.17</u>	<u>93.87 ± 0.63</u>	<u>73.17 ± 0.46</u>
Swin-B	SimMIM	79.09 ± 0.57	86.75 ± 0.96	93.03 ± 0.48	71.99 ± 0.55
	POPAR^{-1}	**80.51 ± 0.15**	**88.25 ± 0.78**	**96.81 ± 0.40**	**73.58 ± 0.18**

(2) POPAR outperforms self-supervised pretrained models across architectures. To demonstrate the effectiveness of representation learning via our proposed framework, we compare POPAR with SoTA CNN-based and transformer-based SSL methods pretrained on medical images. To do so, we evaluate (1) three recent SSL methods with ResNet-50 backbone, including MoCo v2 [5], Barlow Twins [26], and SimSiam [6], and (2) SimMIM [25], which has shown superior performance over other transformer-based SSL methods in both vision [25] and medical (refer to Table 2) tasks, with ViT-B and Swin-B backbones. All models are pretrained on ChestX-ray14 dataset. As shown in Table 3, even POPAR^{-1}, a downgraded POPAR, yields significantly better performance compared with three SSL methods with ResNet-50 backbone in all target tasks. Moreover, even POPAR^{-1} and POPAR^{-3} (two downgraded POPAR models) outperform SimMIM in all target tasks across Swin-B and ViT-B backbones. These results demonstrate that POPAR models provide more useful representations for various medical imaging tasks.

Table 4. POPAR models outperform fully-supervised pretrained models on ImageNet and ChestX-ray14 datasets in three target tasks across architectures. The best methods are bolded while the second best are underlined. Transfer learning is inapplicable when pretraining and target tasks are the same, denoted by "–".

Backbone	Initialization	ChestX-ray14	CheXpert	ShenZhen	RSNA Pneumonia
ResNet-50	Random	80.40 ± 0.05	86.60 ± 0.17	90.49 ± 1.16	70.00 ± 0.50
	ImageNet-1K	$\mathbf{81.70 \pm 0.15}$	87.17 ± 0.22	94.96 ± 1.19	73.04 ± 0.35
	ChestX-ray14	–	87.40 ± 0.26	96.32 ± 0.65	71.64 ± 0.37
ViT-B	Random	70.84 ± 0.19	80.78 ± 0.13	84.46 ± 1.65	66.59 ± 0.39
	ImageNet-21K	77.55 ± 1.82	83.32 ± 0.69	91.85 ± 3.40	71.50 ± 0.52
	ChestX-ray14	–	84.37 ± 0.42	91.23 ± 0.81	66.96 ± 0.24
	POPAR^{-3}	79.58 ± 0.13	87.86 ± 0.17	93.87 ± 0.63	73.17 ± 0.46
Swin-B	Random	74.29 ± 0.41	85.78 ± 0.01	85.83 ± 3.68	70.02 ± 0.42
	ImageNet-21K	81.32 ± 0.19	87.94 ± 0.36	94.23 ± 0.81	73.15 ± 0.61
	ChestX-ray14	–	87.22 ± 0.22	91.35 ± 0.93	70.67 ± 0.18
	POPAR^{-1}	80.51 ± 0.15	$\mathbf{88.25 \pm 0.78}$	$\mathbf{96.81 \pm 0.40}$	$\mathbf{73.58 \pm 0.18}$

Discussion: By integrating Table 1, 2 and 3, based on the reasoning detailed in the supplementary material (Section B.3), we may infer that POPAR would outperform all baseline approaches if they were pretrained with NIH ChestX-rays14.

(3) POPAR outperforms fully-supervised pretrained models across architectures. We compare POPAR models, which are pretrained on unlabeled images of ChestX-ray14 dataset, with fully-supervised pretrained models on ImageNet and ChestX-ray14 across three architectures: ResNet-50, ViT-B, and Swin-B. We use existing supervised ImageNet models, with CNN and transformer backbones pretrained on ImageNet-1K and ImageNet-21K datasets, respectively. As shown in Table 4, POPAR models provide superior performance over both supervised ImageNet and ChestX-ray14 models across architectures in three target tasks. In particular, the downgraded POPAR models with ViT-B and Swin-B backbones outperform corresponding supervised baselines with the same backbone in all and three target tasks, respectively. Moreover, POPAR^{-1} outperforms supervised models with ResNet-50 backbone in three target tasks. In summary, these results demonstrate that POPAR provides more generic features for various medical imaging tasks.

5 Ablation Study

Impact of Input Resolutions. We evaluate POPAR with ViT-B and Swin-B backbones using four different pretraining and finetuning image resolutions. As shown in Table 1, comparing POPAR^{-1} with POPAR^{-2} indicates that a larger number of shufflable patches provides a larger performance gain on all target tasks. Moreover, with the same number of shufflable patches, POPAR^{-1} with a Swin-B backbone provides superior performance compared with POPAR^{-3} with a ViT-B backbone; as a result, the Swin transformer is the most suggested POPAR backbone. Lastly, POPAR pretrained and finetuned with 448×448 resolution, denoted by POPAR in Table 1, suggests the SoTA performance on all four target tasks. It indicates that the higher input resolution is preferred for all four MIA tasks studied in this paper, since higher resolution provides more detailed anatomical information, thus enhancing the performance of all MIA target tasks.

Pretraining Tasks. POPAR seamlessly combines two tasks: patch order prediction and patch appearance recovery. As shown in our supplementary material (Section C and Table 5), they can be further broken down into three individual sub-tasks: (a) patch order classification, denoted by $\mathcal{T}_{poc}$; (b) misplaced patch appearance recovery, denoted by $\mathcal{T}_{mpr}$; and (c) Models Genesis [27] transformed image restoration, denoted by $\mathcal{T}_{mgr}$. We evaluate the effectiveness of different POPAR pretraining subtasks on the ViT-B backbone. Compared with the Models Genesis [27] transformed image restoration, the patch order prediction task provides a significant performance boost on most target tasks. Furthermore, the combination of the misplaced patch appearance recovery task and the patch order classification task provides an on-par or less performance increment on four target tasks. Finally, we demonstrate that POPAR pretrained with all subtasks provides the highest performance boost.

6 Conclusion

We propose POPAR, a novel transformer-based SSL framework for MIA tasks. POPAR integrates patch order prediction and appearance recovery, capturing not only high-level relationships among anatomical structures but also fine-grained details from medical images. As our future work, we will extend POPAR to 3D and cover target segmentation tasks.

Acknowledgments. This research has been supported in part by ASU and Mayo Clinic through a Seed Grant and an Innovation Grant, and in part by the NIH under Award Number R01HL128785. The content is solely the responsibility of the authors and does not necessarily represent the official views of the NIH. This work has utilized the GPUs provided in part by the ASU Research Computing and in part by the Extreme Science and Engineering Discovery Environment (XSEDE) funded by the National Science Foundation (NSF) under grant numbers: ACI-1548562, ACI-1928147, and ACI-2005632. The content of this paper is covered by patents pending.

References

1. RSNA pneumonia detection challenge (2018). https://www.kaggle.com/c/rsna-pneumonia-detection-challenge
2. Bao, H., Dong, L., Wei, F.: BEiT: BERT pre-training of image transformers. arXiv preprint arXiv:2106.08254 (2021)
3. Brown, T., et al.: Language models are few-shot learners. In: Advances in Neural Information Processing Systems, vol. 33, pp. 1877–1901 (2020)
4. Caron, M., et al.: Emerging properties in self-supervised vision transformers. In: Proceedings of the IEEE/CVF International Conference on Computer Vision, pp. 9650–9660 (2021)
5. Chen, X., Fan, H., Girshick, R., He, K.: Improved baselines with momentum contrastive learning. arXiv preprint arXiv:2003.04297 (2020)
6. Chen, X., He, K.: Exploring simple Siamese representation learning. In: Proceedings of the IEEE/CVF Conference on Computer Vision and Pattern Recognition, pp. 15750–15758 (2021)
7. Chen, X., Xie, S., He, K.: An empirical study of training self-supervised vision transformers. In: Proceedings of the IEEE/CVF International Conference on Computer Vision, pp. 9640–9649 (2021)
8. Deng, J., Dong, W., Socher, R., Li, L.J., Li, K., Fei-Fei, L.: ImageNet: a large-scale hierarchical image database. In: 2009 IEEE Conference on Computer Vision and Pattern Recognition, pp. 248–255. IEEE (2009)
9. Devlin, J., Chang, M.W., Lee, K., Toutanova, K.: BERT: pre-training of deep bidirectional transformers for language understanding. arXiv preprint arXiv:1810.04805 (2018)
10. Doersch, C., Gupta, A., Efros, A.A.: Unsupervised visual representation learning by context prediction. In: Proceedings of the IEEE International Conference on Computer Vision, pp. 1422–1430 (2015)
11. Dosovitskiy, A., et al.: An image is worth 16x16 words: transformers for image recognition at scale. arXiv preprint arXiv:2010.11929 (2020)
12. Haghighi, F., Hosseinzadeh Taher, M.R., Gotway, M.B., Liang, J.: DiRA: discriminative, restorative, and adversarial learning for self-supervised medical image analysis. In: Proceedings of the IEEE/CVF Conference on Computer Vision and Pattern Recognition (CVPR), pp. 20824–20834 (2022)

13. Haghighi, F., Hosseinzadeh Taher, M.R., Zhou, Z., Gotway, M.B., Liang, J.: Learning semantics-enriched representation via self-discovery, self-classification, and self-restoration. In: Martel, A.L., et al. (eds.) MICCAI 2020. LNCS, vol. 12261, pp. 137–147. Springer, Cham (2020). https://doi.org/10.1007/978-3-030-59710-8_14
14. Haghighi, F., Taher, M.R.H., Zhou, Z., Gotway, M.B., Liang, J.: Transferable visual words: exploiting the semantics of anatomical patterns for self-supervised learning. IEEE Trans. Med. Imaging **40**(10), 2857–2868 (2021)
15. He, K., Chen, X., Xie, S., Li, Y., Dollár, P., Girshick, R.: Masked autoencoders are scalable vision learners. arXiv preprint arXiv:2111.06377 (2021)
16. Hosseinzadeh Taher, M.R., Haghighi, F., Feng, R., Gotway, M.B., Liang, J.: A systematic benchmarking analysis of transfer learning for medical image analysis. In: Albarqouni, S., et al. (eds.) DART/FAIR -2021. LNCS, vol. 12968, pp. 3–13. Springer, Cham (2021). https://doi.org/10.1007/978-3-030-87722-4_1
17. Irvin, J., et al.: CheXpert: a large chest radiograph dataset with uncertainty labels and expert comparison. In: Proceedings of the AAAI Conference on Artificial Intelligence, vol. 33, pp. 590–597 (2019)
18. Jaeger, S., Candemir, S., Antani, S., Wáng, Y.X.J., Lu, P.X., Thoma, G.: Two public chest X-ray datasets for computer-aided screening of pulmonary diseases. Quant. Imaging Med. Surg. **4**(6), 475 (2014)
19. Jing, L., Tian, Y.: Self-supervised visual feature learning with deep neural networks: a survey. IEEE Trans. Pattern Anal. Mach. Intell. **43**(11), 4037–4058 (2020)
20. Noroozi, M., Favaro, P.: Unsupervised learning of visual representations by solving jigsaw puzzles. In: Leibe, B., Matas, J., Sebe, N., Welling, M. (eds.) ECCV 2016. LNCS, vol. 9910, pp. 69–84. Springer, Cham (2016). https://doi.org/10.1007/978-3-319-46466-4_5
21. Schwichtenberg, J.: Physics from Symmetry. Springer, Cham (2015). https://doi.org/10.1007/978-3-319-66631-0
22. Taher, M.R.H., Haghighi, F., Gotway, M.B., Liang, J.: CAiD: context-aware instance discrimination for self-supervised learning in medical imaging. arXiv:2204.07344 (2022)
23. Tajbakhsh, N., Roth, H., Terzopoulos, D., Liang, J.: Guest editorial annotation-efficient deep learning: the holy grail of medical imaging. IEEE Trans. Med. Imaging **40**(10), 2526–2533 (2021)
24. Wang, X., Peng, Y., Lu, L., Lu, Z., Bagheri, M., Summers, R.M.: ChestX-ray8: hospital-scale chest X-ray database and benchmarks on weakly-supervised classification and localization of common thorax diseases. In: Proceedings of the IEEE Conference on Computer Vision and Pattern Recognition, pp. 2097–2106 (2017)
25. Xie, Z., et al.: SimMIM: a simple framework for masked image modeling. arXiv preprint arXiv:2111.09886 (2021)
26. Zbontar, J., Jing, L., Misra, I., LeCun, Y., Deny, S.: Barlow twins: self-supervised learning via redundancy reduction. In: International Conference on Machine Learning, pp. 12310–12320. PMLR (2021)
27. Zhou, Z., Sodha, V., Pang, J., Gotway, M.B., Liang, J.: Models genesis. Med. Image Anal. **67**, 101840 (2021)
28. Zhuang, X., Li, Y., Hu, Y., Ma, K., Yang, Y., Zheng, Y.: Self-supervised feature learning for 3D medical images by playing a Rubik's cube. In: Shen, D., et al. (eds.) MICCAI 2019. LNCS, vol. 11767, pp. 420–428. Springer, Cham (2019). https://doi.org/10.1007/978-3-030-32251-9_46

Feather-Light Fourier Domain Adaptation in Magnetic Resonance Imaging

Ivan Zakazov[1,2], Vladimir Shaposhnikov[1,2], Iaroslav Bespalov[1,2], and Dmitry V. Dylov[2(✉)]

[1] Philips Research, Moscow, Russia
[2] Skolkovo Institute of Science and Technology, Moscow, Russia
D.Dylov@skoltech.ru

Abstract. Generalizability of deep learning models may be severely affected by the difference in the distributions of the train (*source domain*) and the test (*target domain*) sets, *e.g.*, when the sets are produced by different hardware. As a consequence of this *domain shift*, a certain model might perform well on data from one clinic, and then fail when deployed in another. We propose a very light and transparent approach to perform *test-time domain adaptation*. The idea is to substitute the *target* low-frequency Fourier space components that are deemed to reflect the style of an image. To maximize the performance, we implement the *"optimal style donor"* selection technique, and use a number of *source* data points for altering a single *target* scan appearance (*Multi-Source Transferring*). We study the effect of severity of domain shift on the performance of the method, and show that our *training-free* approach reaches the state-of-the-art level of complicated deep domain adaptation models. The code for our experiments is released (https://github.com/kechua/Feather-Light-Fourier-Domain-Adaptation/).

Keywords: Domain adaptation · MRI · Fourier Domain Adaptation · Test-time domain adaptation

1 Introduction

Magnetic Resonance Imaging (MRI) has become an irreplaceable tool in healthcare thanks to its capacity to produce high-resolution scans without ionizing radiation. The widespread use of the modality has helped to accumulate large volumes of miscellaneous imaging data, which have been fueling development of machine- and deep-learning methods, aiming to mimic diagnostic decisions. However, the real-life deployment of these methods is often hindered by an issue known as *Domain Shift*, which originates from a possible difference in train

I. Zakazov and V. Shaposhnikov—Equal contribution.

Supplementary Information The online version contains supplementary material available at https://doi.org/10.1007/978-3-031-16852-9_9.

K. Kamnitsas et al. (Eds.): DART 2022, LNCS 13542, pp. 88–97, 2022.
https://doi.org/10.1007/978-3-031-16852-9_9

(*source*) and test (*target*) distributions. This difference might occur whenever *source* and *target* datasets are acquired with different machines or research protocols, and entails a need for proper *Domain Adaptation* (DA) [4].

To this end, modern DA in medical imaging includes a plethora of shallow and deep models [4]. Existing shallow DA methods are somewhat rudimentary, requiring human-engineered features [21,22], and generally lagging in performance, while the deep ones are often heavy, slow, and barely interpretable (albeit accurate) [8,15,24].

One trait shared by all of the aforementioned methods is that they operate in the *image space*. It is only recently that the community has started to realize the potential of operating in k-space (also referred to as *Fourier space* or *spectrum* of an image) for tackling domain shift in MRI data with [10,11] being the only two studies we were able to find. This is surprising given that MRI is a modality that yields k-space data representation by design.

In k-space, specific spectral components are responsible for different properties of an image, *e.g.*, the high frequency components accentuate the edges and enhance details, while the low ones control contrast and the large-scale content. Besides, it is known that the semantic information is mostly stored in the phase component of the spectrum. This suggests an efficient strategy for tackling DA via "mixing" the *source* and the *target* spectra. The purpose of this "mixing" (*Fourier Domain Adaptation* or *FDA*) is to transfer the style while **preserving the patient-specific content**, thereby compensating for the DA shift.

The idea is borrowed from the natural image domain [26] and adapted to MRI volumes, entailing a new *"optimal style donor"* selection module and a *multi-source* transferring routine (we use multiple *source* images when transferring the style to a given *target* image to further improve the performance).

We call our method *feather-light* because, unlike the modern go-to approaches, it **does not involve any training**, as we simply transfer the k-space components, characterizing the style of the *source* domain, to a *target* scan during the test time. Despite its simplicity, the method performs on par with complicated deep DA models, such as those based on Generative Adversarial Networks (GANs) [23]. Notably, the proposed method is also *interpretable* as it directly shows which style-carrying source frequencies alleviate the domain shift.

2 Related Work

A large part of Deep Domain Adaptation methods could be split into feature-level [2,13] and image-level approaches. Among the image-level ones, the majority exploit the idea of GANs [7,10,23] for eradicating the difference in distribution between images from various domains. GANs, however, are difficult to train, lack explainability and might produce undesirable artefacts, which pose an even greater problem in the medical imaging context

Fourier Domain Adaptation (**FDA**) [26] provides a feasible alternative to GANs, as image-to-image translation, performed via low-frequency spectra components swap (amplitudes only), is simple, predictable and yet yields SOTA-level

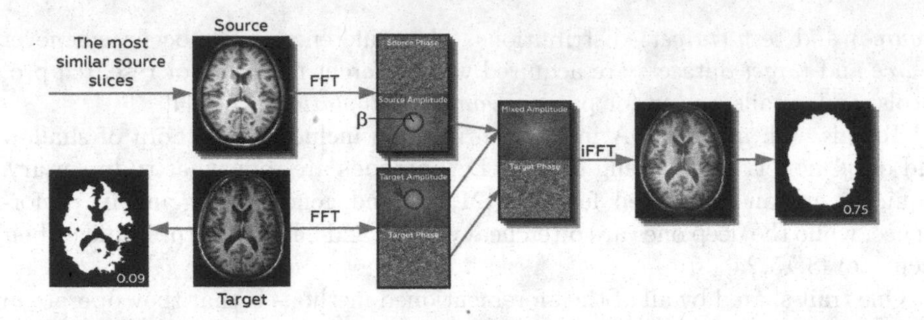

Fig. 1. Fourier Domain Adaptation (FDA) for the Brain Segmentation task.

results on the natural images. This method has been adapted for the medical imaging, with the earliest application being mitigation of domain shift, appearing in the synthetic ultrasound images [17]. In [10] the authors applied FDA-based augmentation technique for the cardiac MRI segmentation, with the novelty being swapping both amplitudes and phases, which apparently is not very stable and may lead to changes in the image semantics [25]. [11] applies FDA to federated learning in order to generate images exhibiting distribution characteristic of other clients, while [1] uses FDA as a proof-of-concept tool for obtaining "poisoned" images, challenging for neural networks.

In [12] the authors solve automatic polyp detection task via combining feature-level adaptation with FDA, while further improving the FDA component with sampling "matching" source and target image pairs. The closer a target image is to the source one in terms of cosine similarity of their deep ResNet-50 features, the greater is the "match" probability. We note that adding an additional deep model to the pipeline makes it more complicated, while we strive for simplicity.

One dismissed idea in the FDA area is the one we propose to denote *multi-source transfer*, *i.e.*, performing a number of k-space components swaps with a single *target* image and multiple *source* images followed by averaging of the down-stream task predictions for various versions of the changed *target*.

3 Method

We base our approach on the Fourier Domain Adaptation technique [26] and summarize it in Fig. 1. This method consists of swapping the low-frequency amplitudes of an image spectrum with those of another image, the style of which should be borrowed. As amplitudes of the low-frequency spectrum components are mostly related to the low-level image characteristics, defining the style, this procedure is expected to align the *source* and the *target* distributions, thus compensating for the *domain shift* between them.

While in [26], the *source* data is transferred to the *target* style, and the deep neural network is then trained on the $D^{s \to t}$ dataset, we note that in the clinical setting it would mean re-training the model for each new *target* domain (e.g., a new clinic), which might complicate certification and clinical deployment. Moreover, the *target* data (*e.g.*, data from a hospital we need to adjust the model for) may appear to be scarce, which limits capabilities of the self-supervised training on *target*, another component of the original method.

Instead of $D^{s \to t}$, we focus on the $D^{t \to s}$ adaptation with a single *source model* used for various *targets*, which are transferred into the *source* style during the test time. Mathematically speaking, we carry out the following procedure: $t_{new} = f_{FDA}(s_i, t) = \mathcal{F}^{-1}\left(\left[M_\beta \circ \mathcal{F}^A(s_i) + (1 - M_\beta) \circ \mathcal{F}^A(t), \mathcal{F}^P(t)\right]\right)$.

The phase component of a spectrum remains intact, while the source style is injected with the low-frequency amplitudes we "cut out" with M_β (Fig. 1).

As there is no additional training required, this setting is much more light-weight than the original one, but it is also more challenging in terms of reaching the optimal performance. To this end, we improve the method in the following ways:

- We propose to carry out a multitude of $t \to s_i$ swaps (Multi-Source Transfer) with the final result for slice t calculated as $\sum_{i=1}^{n_{MST}} net(f_{FDA}(s_i, t))/n_{MST}$, where $f_{FDA}(s_i, t)$ is the FDA procedure, performed on s_i and t; n_{MST} reflects the number of *source* slices used for the style transfer (after preliminary experiments we set $n_{MST} = 7$)
- We design an approach for picking the optimal *source* slices.

The intuition behind the latter feature is that as swapping the spectral components inevitably leads to artefacts, we should minimize this detrimental effect by choosing *source* and *target* which are as close as possible in terms of their semantics. To do so, we assess the "closeness" with the Spectral Residual Similarity (SR-SIM) semantic similarity measure [27] (Fig. 2).

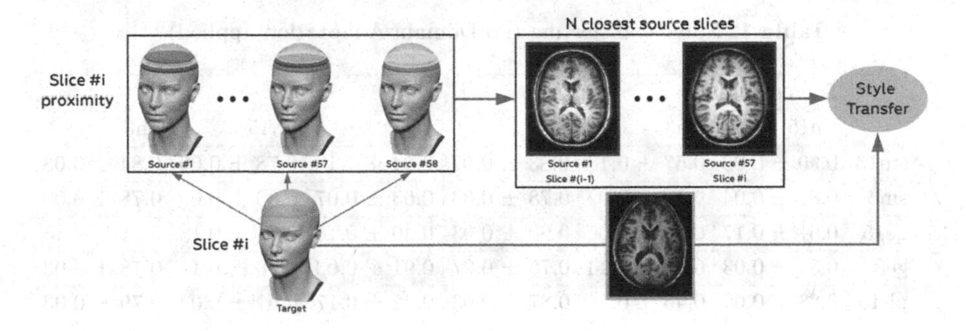

Fig. 2. Multi-Source Transfer (MST) + SR-SIM source choice in the **2.5D** fashion.

We consider several approaches to the optimal *source "style donor"* search for the slices, belonging to the *target* scan. Firstly, we may average the similarity score between the corresponding slices in 2 scans, thus obtaining the scan-to-scan similarity score ($sim(s, t) = \sum_{i=1}^{n_{slices}} sim(s_i, t_i)/n_{slices}$). We then select for Domain Adaptation n_{MST} most similar *source* scans (**3D** similarity).

We note that choosing the "style donors" on the scan level might introduce unnecessary constraint. Alternatively, *source* "style donors" for various t_i slices of the target scan t may come from different *source* scans. In this case, for $i \in (1, n_{slices})$ we look for the slices closest to t_i among $(s_i^1, s_i^2, ..., s_i^{n_{scans}})$ (**2D**). A natural extension to this approach is broadening this set to $(s_{i-m}^1, s_{i-m+1}^1, ..., s_i^1, s_{i+1}^1, s_{i+m}^1, s_{i-m}^2, ..., s_{i+m}^{n_{scans}})$, which we refer to as **2.5D** (Fig. 2). We set $m = 2$.

4 Experiments

4.1 Technical Details

We conduct all the experiments on a public brain MR dataset called CC359 [20], which is formed of 359 scans and various masks, among which are the *brain segmentation* masks. The scans are produced by one of 6 MRI machines (Siemens, Philips, GE; 1.5T or 3T each), and thus fall into one of 6 domains of approximately equal size (60 or 59 scans). We perform affine registration of all the scans to MNI152 template using the FSL software [5,6], and subsequently normalize voxel intensities to $[0, 1]$. We use the *Surface Dice Score* [14] as it appears to be a more reliable indicator of the brain segmentation quality than the standard Dice Score [19].

We solve the brain segmentation task with 2D U-Net with residual blocks, which we train for 100 epochs (100 iterations per epoch), using SGD optimizer with Nesterov momentum of 0.9, combination of BCE and dice losses (weighted with 0.4 and 0.6 coefficients), and learning rate of 10^{-3}, reduced to 10^{-4} at epoch 80. We train the networks on $(256, 256)$ crops grouped in batches of 16 samples. The crops are sampled randomly at each iteration.

Table 1. Naive transferring (no Domain Adaptation applied).

		Source domains					
		sm15	sm3	ge15	ge3	ph15	ph3
Target domains	sm15	0.90 ± 0.03	0.57 ± 0.18	0.83 ± 0.07	0.54 ± 0.18	0.78 ± 0.09	0.84 ± 0.03
	sm3	0.81 ± 0.04	0.90 ± 0.02	0.78 ± 0.03	0.63 ± 0.07	0.80 ± 0.05	0.78 ± 0.03
	ge15	0.61 ± 0.17	0.11 ± 0.06	0.90 ± 0.03	0.40 ± 0.16	0.51 ± 0.18	0.67 ± 0.15
	ge3	0.84 ± 0.03	0.44 ± 0.14	0.78 ± 0.07	0.91 ± 0.03	0.76 ± 0.1	0.78 ± 0.03
	ph15	0.83 ± 0.06	0.45 ± 0.1	0.87 ± 0.03	0.42 ± 0.17	0.91 ± 0.03	0.79 ± 0.03
	ph3	0.74 ± 0.12	0.40 ± 0.12	0.62 ± 0.12	0.39 ± 0.12	0.56 ± 0.12	0.88 ± 0.04

We follow the methodology of the original Fourier Domain Adaptation (FDA) paper [26] with respect to the k-space swapping technique, with the notable difference of using the circular crop instead of the rectangular one, which is to take into account the radial symmetry of the spectrum components amplitudes.

4.2 Naive Model Transfer

Firstly, we consider a simple case of *no Domain Adaptation* applied. In this regard, we train 6 *base models* on the corresponding domains, designating all but 2 *source* scans for training (these 2 are to ensure reaching the loss plateau when training). We then transfer these source-trained models to *unseen domains*, thus considering 30 source-target pairs. We calculate each transferred model performance on the *target test set* of 10 images. Besides, we also use 3-fold cross-validation to assess the model performance on the domain it was trained on.

As may be seen from Fig. 1, the magnitude of *Domain Shift*, *i.e.*, the performance variability between the transferred model and the one which was initially trained on some domain changes significantly between the source-target pairs. As conducting subsequent experiments on all 30 source-target pairs is computationally prohibitive, we decide to concentrate on 3 clusters, representing **severe** domain shift, **medium** domain shift and the **subtle** one. We sort the source-target pairs by the metric decline magnitude, and pick 2 pairs per cluster from the top, bottom, and middle of this sorted list.

4.3 Choosing the Optimal β

One of the most important FDA design choices is choosing the size of the swapping window β. Specifically for this purpose we designate another 10 *target* scans per source-target pair, on which the grid search over various β values is performed. We consider 2 strategies of devising the optimal β, which correspond to 2 actual clinical set-ups:

- **Optimal per Pair**. Picking the β, which proved to be the optimal one for each source-target pair. This set-up is motivated by the scenario, in which at least some target domain data (*e.g.*, data coming from a new clinical center) is available and labelled, and thus may be used for setting the optimal β, peculiar to this source-target pair
- **Averaged Optimal**. Picking the β, based on the grid-search results, averaged over all pairs, which corresponds to a broader scenario of setting a single "standard" beta for all the pairs

Table 2. Comparison of the performance of various proposed methods with the baselines. *Style* is for StyleGAN [9], *Cycle* is for CycleGAN [23], *Fast* is for artistic stylization network [3].

		No DA	Baselines				β: aver. optimal			β: opt. per pair		
			StyleSegor	Cycle	Style	Fast	3D	2.5D	2D	3D	2.5D	2D
Shift severity	Severe #1	0.11	0.11	0.50	0.46	0.15	0.57	0.57	0.57	0.57	0.58	**0.59**
	Severe #2	0.39	0.46	**0.64**	0.58	0.12	0.50	0.48	0.47	0.51	0.48	0.48
	Medium #1	0.67	0.66	0.70	0.64	0.15	0.72	0.73	0.74	**0.77**	**0.77**	**0.77**
	Medium #2	0.74	0.69	0.69	0.61	0.11	0.72	0.69	0.68	**0.75**	0.74	0.73
	Subtle #1	0.84	**0.85**	0.60	0.41	0.17	0.81	0.81	0.82	0.83	0.83	0.84
	Subtle #2	**0.87**	0.82	0.46	0.55	0.11	0.85	0.83	0.84	**0.87**	0.86	0.86
	Average	0.60	0.60	0.60	0.54	0.14	0.7	0.68	0.69	**0.72**	0.71	0.71

Table 3. The ablation study.

		β: averaged optimal			β: optimal per pair		
		SR-SIM+	MST	None	SR-SIM+	MST	None
Shift severity	Severe #1	**0.57**	0.41	0.40	**0.59**	0.57	0.53
	Severe #2	**0.47**	0.42	0.41	**0.48**	0.41	0.41
	Medium #1	0.74	**0.78**	0.76	0.77	**0.78**	0.76
	Medium #2	**0.68**	0.69	0.65	**0.73**	0.69	0.65
	Subtle #1	0.82	**0.85**	0.82	0.84	**0.85**	0.82
	Subtle #2	0.84	**0.85**	0.84	0.86	**0.85**	0.84
	Average	**0.69**	0.67	0.65	**0.71**	0.69	0.67

4.4 Results and Discussion

As was discussed in Sect. 3, we consider 3 approaches to picking the "best" source slices, which we denote **2D**, **2.5D**, and **3D**. Furthermore, in line with 4.3 we devise β from the *target* validation set by means of either *global averaging* or *averaging per pair*. The corresponding results are presented in Table 2 in comparison with the SOTA-level baselines of CycleGAN [23], StyleGAN [10] and Style-Segor [13]. We also consider another light-weight baseline [3].

We perform the ablation study (Table 3), comparing (a) SR-SIM chosen sources + Multi-source transfer (MST) (b) Multi-source transfer (MST) (c) A "simple" swap.

2D *vs.* 2.5D *vs.* 3D. Interestingly, no significant difference could be observed between various SR-SIM-based source choice approaches. Subsequently, we concentrate on the **3D** approach, as it appears to be both more intuitive and marginally better than the others

β: Averaged Optimal *vs.* β: Optimal per Pair. Picking β on *target* Validation set in a pair-wise fashion gives only a minor advantage over devising it from the averaged *target* validation $sDice(\beta)$ curve. Besides, the latter set-up does not require adjusting β for a particular pair on a labelled subset of target data, and thus is more relevant in the clinical practice. Therefore, from now on we concentrate on the *Averaged Optimal* results analysis.

Our method (3D; β: Averaged Optimal) *vs.* Baselines. While our method is outperformed by **CycleGAN** on *severe #2* pair and by **StyleSegor** on *subtle #1* pair, it is the only one demonstrating good performance across all the data shift magnitude range, since **GANs** fail to preserve even the "naive" swap quality (**no DA**) in case of low domain shift and **StyleSegor** is barely improving the score in case of strong domain shift. Fast artistic image stylization [3], another light-weight method we consider as a baseline, does not demonstrate sufficiently good performance.

Ablation Studies . As could be seen in Table 3, both introducing Multi-Source Transfer and combining it with the SR-SIM-based source choice improves the

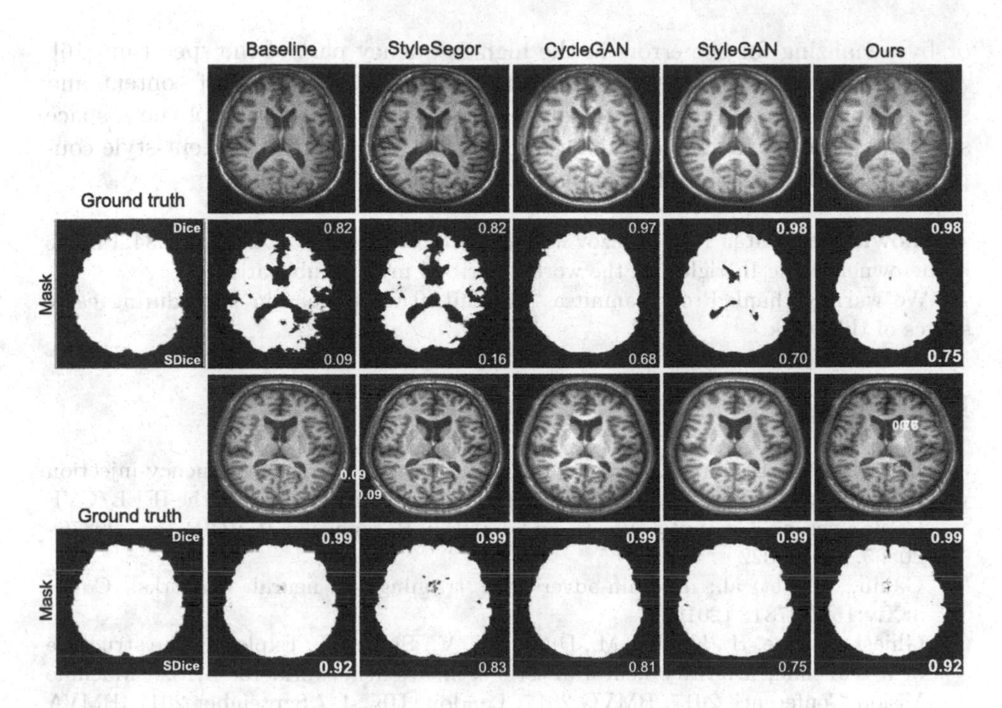

Fig. 3. Visual comparison of various approaches. In this particular case, we set $\beta = 0.03$

score on average with the positive effect of the "smart" source choice substantial for the instances of severe domain shift.

In Fig. 3, we visually compare our approach with the baselines, considering the cases of severe (top) and subtle (bottom) domain shift. For illustration purposes, we apply the method to the middle-positioned slices. Notably, in case of severe domain shift GANs alter the appearance much more significantly, which might explain the decreasing score.

5 Conclusion

We present a novel Fourier-based Domain Adaptation method, which requires neither any training, nor incorporating any additional deep components into the pipeline. We consider various domain shift severity scenarios, and show that our method performs consistently across all of them, outperforming SOTA-level GANs in case of *subtle* domain shift. We note that the simplicity achieved ensures better explainability, and envision easier certification, as we avoid modifying the deep model in any way, but rather adapt an incoming image in a strictly defined fashion.

A limitation of this study is the blunt selection of the k-space low-frequency window, which could be improved by engaging intelligent search for the style-bearing spectrum components, such as presented in [18] for the supervised case, or by penalizing for the errors in the high-frequency part of the spectrum [16]. Another fundamental assumption we make is the separability of content and style, which is known to be true only partially [2]. Optimization of the k-space swapping pattern along with taking into account the intrinsic content-style coupling will be the subject of future work.

Acknowledgements. Ivan Zakazov was supported by RSF grant 20-71-10134. Philips is the owner of the IP rights on the work described in this publication.

We warmly thank Prof. Kamnitsas for fruitful discussions in 2021 during early stages of this work.

References

1. Feng, Y., Ma, B., Zhang, J., Zhao, S., Xia, Y., Tao, D.: FIBA: frequency-injection based backdoor attack in medical image analysis. In: Proceedings of the IEEE/CVF Conference on Computer Vision and Pattern Recognition (CVPR), pp. 20876–20885, June 2022
2. Ganin, Y., et al.: Domain-adversarial training of neural networks. CoRR arXiv:1505.07818 (2015)
3. Ghiasi, G., Lee, H., Kudlur, M., Dumoulin, V., Shlens, J.: Exploring the structure of a real-time, arbitrary neural artistic stylization network. In: British Machine Vision Conference 2017, BMVC 2017, London, UK, 4–7 September 2017. BMVA Press (2017)
4. Guan, H., Liu, M.: Domain adaptation for medical image analysis: a survey. IEEE Trans. Biomed. Eng. **69**(3), 1173–1185 (2022)
5. Jenkinson, M., Bannister, P.R., Brady, M., Smith, S.M.: Improved optimization for the robust and accurate linear registration and motion correction of brain images. Neuroimage **17**, 825–841 (2002)
6. Jenkinson, M., Smith, S.: Global optimisation method for robust affine registration of brain images. Med. Image Anal. **5**, 143–56 (2001)
7. Joshi, N., Burlina, P.: AI fairness via domain adaptation. arXiv:2104.01109 (2021)
8. Kamnitsas, K., et al.: Unsupervised domain adaptation in brain lesion segmentation with adversarial networks. In: Niethammer, M., et al. (eds.) IPMI 2017. LNCS, vol. 10265, pp. 597–609. Springer, Cham (2017). https://doi.org/10.1007/978-3-319-59050-9_47
9. Karras, T., Laine, S., Aittala, M., Hellsten, J., Lehtinen, J., Aila, T.: Analyzing and improving the image quality of StyleGAN. In: 2020 IEEE/CVF Conference on Computer Vision and Pattern Recognition (CVPR), pp. 8107–8116 (2020)
10. Kong, F., Shadden, S.C.: A generalizable deep-learning approach for cardiac magnetic resonance image segmentation using image augmentation and attention U-Net. In: Puyol Anton, E., et al. (eds.) STACOM 2020. LNCS, vol. 12592, pp. 287–296. Springer, Cham (2021). https://doi.org/10.1007/978-3-030-68107-4_29
11. Liu, Q., Chen, C., Qin, J., Dou, Q., Heng, P.A.: FedDG: federated domain generalization on medical image segmentation via episodic learning in continuous frequency space. In: The IEEE/CVF Conference on Computer Vision and Pattern Recognition (CVPR) (2021)
12. Liu, X., Guo, X., Liu, Y., Yuan, Y.: Consolidated domain adaptive detection and localization framework for cross-device colonoscopic images. Med. Image Anal. **71**, 102052 (2021)

13. Ma, C., Ji, Z., Gao, M.: Neural style transfer improves 3D cardiovascular MR image segmentation on inconsistent data. In: Shen, D., et al. (eds.) MICCAI 2019. LNCS, vol. 11765, pp. 128–136. Springer, Cham (2019). https://doi.org/10.1007/978-3-030-32245-8_15
14. Nikolov, S., et al.: Deep learning to achieve clinically applicable segmentation of head and neck anatomy for radiotherapy. arXiv preprint arXiv:1809.04430 (2018)
15. Perone, C.S., Ballester, P., Barros, R.C., Cohen-Adad, J.: Unsupervised domain adaptation for medical imaging segmentation with self-ensembling. Neuroimage **194**, 1–11 (2019)
16. Pronina, V., Kokkinos, F., Dylov, D.V., Lefkimmiatis, S.: Microscopy image restoration with deep Wiener-Kolmogorov filters. In: Vedaldi, A., Bischof, H., Brox, T., Frahm, J.-M. (eds.) ECCV 2020. LNCS, vol. 12365, pp. 185–201. Springer, Cham (2020). https://doi.org/10.1007/978-3-030-58565-5_12
17. Sharifzadeh, M., Tehrani, A.K.Z., Benali, H., Rivaz, H.: Ultrasound domain adaptation using frequency domain analysis. arXiv:2109.09969 (2021)
18. Shipitsin, V., Bespalov, I., Dylov, D.V.: GAFL: global adaptive filtering layer for computer vision. arXiv: 2010.01177 (2021)
19. Shirokikh, B., Zakazov, I., Chernyavskiy, A., Fedulova, I., Belyaev, M.: First U-Net layers contain more domain specific information than the last ones. In: Albarqouni, S., et al. (eds.) DART/DCL -2020. LNCS, vol. 12444, pp. 117–126. Springer, Cham (2020). https://doi.org/10.1007/978-3-030-60548-3_12
20. Souza, R., et al.: An open, multi-vendor, multi-field-strength brain MR dataset and analysis of publicly available skull stripping methods agreement. Neuroimage **170**, 482–494 (2018)
21. Wang, J., et al.: Multi-class ASD classification based on functional connectivity and functional correlation tensor via multi-source domain adaptation and multi-view sparse representation. IEEE Trans. Med. Imaging **39**(10), 3137–3147 (2020)
22. Wang, M., Zhang, D., Huang, J., Yap, P.T., Shen, D., Liu, M.: Identifying autism spectrum disorder with multi-site fMRI via low-rank domain adaptation. IEEE Trans. Med. Imaging **39**(3), 644–655 (2020)
23. Welander, P., Karlsson, S., Eklund, A.: Generative adversarial networks for image-to-image translation on multi-contrast MR images - a comparison of CycleGAN and UNIT. CoRR arXiv:1806.07777 (2018)
24. Wollmann, T., Eijkman, C.S., Rohr, K.: Adversarial domain adaptation to improve automatic breast cancer grading in lymph nodes. In: 2018 IEEE 15th International Symposium on Biomedical Imaging (ISBI 2018), pp. 582–585 (2018)
25. Yang, Y., Lao, D., Sundaramoorthi, G., Soatto, S.: Phase consistent ecological domain adaptation, pp. 9008–9017 (2020). https://doi.org/10.1109/CVPR42600.2020.00903
26. Yang, Y., Soatto, S.: FDA: Fourier domain adaptation for semantic segmentation. In: 2020 IEEE/CVF Conference on Computer Vision and Pattern Recognition (CVPR), pp. 4084–4094 (2020)
27. Zhang, L., Li, H.: SR-SIM: a fast and high performance IQA index based on spectral residual. In: 2012 19th IEEE International Conference on Image Processing, pp. 1473–1476 (2012)

Seamless Iterative Semi-supervised Correction of Imperfect Labels in Microscopy Images

Marawan Elbatel(✉), Christina Bornberg, Manasi Kattel, Enrique Almar, Claudio Marrocco, and Alessandro Bria

University of Cassino and Southern Lazio, Cassino, Italy
marawan.elbatel@studentmail.unicas.it

Abstract. In-vitro tests are an alternative to animal testing for the toxicity of medical devices. Detecting cells as a first step, a cell expert evaluates the growth of cells according to cytotoxicity grade under the microscope. Thus, human fatigue plays a role in error making, making the use of deep learning appealing. Due to the high cost of training data annotation, an approach without manual annotation is needed. We propose *Seamless Iterative Semi-Supervised correction of Imperfect labels (SISSI)*, a new method for training object detection models with noisy and missing annotations in a semi-supervised fashion. Our network learns from noisy labels generated with simple image processing algorithms, which are iteratively corrected during self-training. Due to the nature of missing bounding boxes in the pseudo labels, which would negatively affect the training, we propose to train on dynamically generated synthetic-like images using seamless cloning. Our method successfully provides an adaptive early learning correction technique for object detection. The combination of early learning correction that has been applied in classification and semantic segmentation before and synthetic-like image generation proves to be more effective than the usual semi-supervised approach by >15% AP and >20% AR across three different readers. Our code is available at https://github.com/marwankefah/SISSI.

Keywords: Label correction · Cell detection · Semi-supervised object detection

1 Introduction

Testing medical devices with animals have a long tradition according to ISO 10993 [1]. Since 2017 the ISO 10993 has gradually evolved towards implementing

M. Elbatel and C. Bornberg—Co-first authors.

Supplementary Information The online version contains supplementary material available at https://doi.org/10.1007/978-3-031-16852-9_10.

K. Kamnitsas et al. (Eds.): DART 2022, LNCS 13542, pp. 98–107, 2022.
https://doi.org/10.1007/978-3-031-16852-9_10

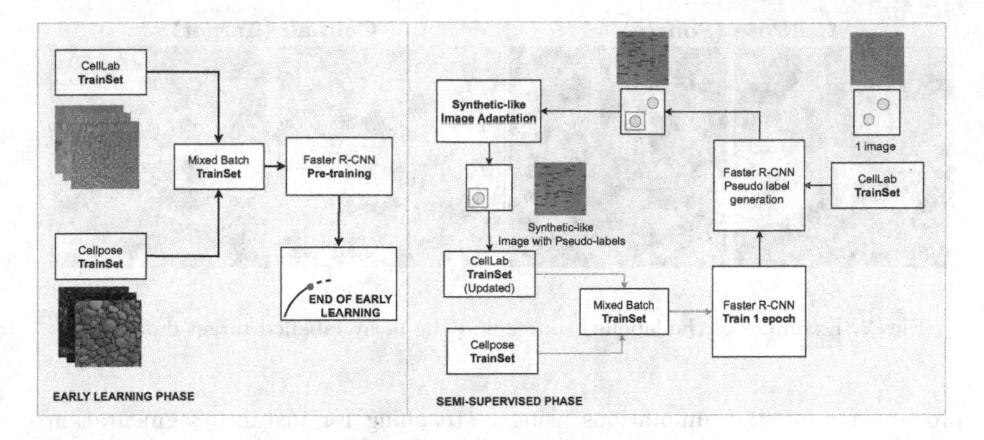

Fig. 1. Overall scheme of SISSI framework.

alternative test methods. One of the in-vitro methods is the testing of cytotoxicity, described in the ISO 10993-5 [3]. Cell experts analyze cell growth of a fibroblast cell line such as L929 with the help of a microscope. The acceptance criteria for medical devices is 50% of dead cells (grade 2 criteria). If there are more than 50% dead cells, the medical device is not allowed to enter the market.

In this context, deep learning can serve as a second opinion since human error in the workplace is costly and dependent on the level of fatigue; the greater the level of fatigue, the higher the risk of errors occurring. Especially in the borderline cases of grade 2, the cell expert needs to be able to obtain a second opinion that is independent of human fatigue. Deep learning has shown substantial benefits in different life science and pharma applications such as chemo-informatics, computational genomics, and biomedical imaging such as cell segmentation [12] and seems to be a promising supplement to cytotoxicity grading. In the first instance, cells need to be detected, and in future work, an intuitive way of classifying cells into dead or alive needs to be found.

When dealing with imperfect datasets, problems including (partly) missing, inaccurate, or wrong labels arise. To handle imperfect datasets in object detection/segmentation tasks, one can leverage unlabelled (self/semi-supervised) or external labelled (transfer learning) data, regularise training, learn with class labels, and revisit loss functions (sparse/noisy labels) [15].

2 Related Works

Previous work has studied imperfect datasets, including semantic segmentation, instance segmentation, and object detection [4,18,19]. [18] propose a pipeline for semantic semi-supervised segmentation that separates pixels of a pseudo labelled image into reliable and unreliable. [6] propose Adaptive Early Learning Correction (ADELE) for semantic segmentation, with a supervised early-learning phase and subsequently a label correction phase. [8] propose a label mining

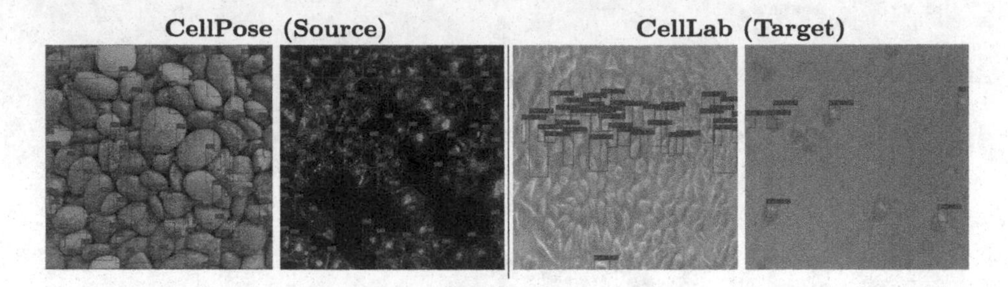

Fig. 2. Examples of the labelled source and the noisy labelled target datasets.

pipeline for missing annotations using co-teaching for instance segmentation. [19] propose to generate masks with the Circle Hough Transform (CHT) and iteratively create pseudo labels with self-training for images where CHT failed. [20] propose to use a background calibration loss inspired by focal loss for object detection with missing annotations. [4] propose only annotating one instance per category in an image and iteratively generating pseudo-labels. [2] propose an object detector to handle noisy labels, masking the negative sample loss in the box predictor to avoid the harm of false-negative labels.

Though advances in dealing with imperfect datasets have been made, the problem of dealing with datasets having partly missing labels that are additionally noisy in object detection tasks remains.

We propose SISSI (Seamless Iterative Semi-Supervised correction of Imperfect labels) for training object detection models with noisy and missing labels in a semi-supervised fashion, see Fig. 1. We perform several experiments with mixed-batch training, self-training with iterative label correction, synthetic-like image generation, and altering the starting point of self-training (ADELE vs. validation loss).

3 Materials and Methods

3.1 Datasets

Microscopy images of fibroblast (L929) were acquired using a Nikon Eclipse TS 100 microscope and the OPTOCAM-I camera. This trainset (**CellLab** dataset) consists of 224 images, and their noisy annotations are generated with simple image processing pipelines such as Circle Hough Transform, Watershed, and Edge Detection. A detailed description of the initial weak label generation is shown in Fig. 5 in Appendix A. The CellLab testset consists of five images (640×480) annotated by three cell experts. Three readers annotated five images independently, resulting in (reader 1) 552, (reader 2) 565, and (reader 3) 477 annotated cells for the five images. In order to perform domain adaptation and enhance our weak and noisy labelled CellLab dataset, we use the labelled **Cellpose** [14] dataset. It consists of a large variety of fluorescent markers and image modalities, as well as natural images that can be segmented into repetitive structures/blobs. The Cellpose dataset is used for training (45,215 cells on 539 images) and validation (7,195 cells on 68 images). We extract bounding boxes from the segmentation masks for our detection task. We show examples of both datasets in Fig. 2.

3.2 Overall Framework

SISSI integrates a range of image processing and deep learning methods to make iterative label correction possible.

The **early learning phase** consists of a mixed-batch training combining the CellLab and Cellpose training datasets. We train the Faster R-CNN model in a supervised fashion with a Balanced Gradient Contribution [11], mixed-batch training, of target dataset with initial noisy annotations and source dataset until a memorisation phase on the noisy annotations is reached. We determine the end of early learning with a deceleration point based on the AP_{50} curve between the weak ground truth of the CellLab dataset and the model output.

In the following **semi-supervised phase** for each cycle, first, we apply label correction, followed by mixed-batch training with the pseudo labels and synthetic-like images (excluding undetected cells) of the CellLab dataset combined with the original Cellpose dataset. Pseudo-label generation uses test-time augmentation and weighted boxes fusion to generate confident bounding boxes. Since some cells are not detected, their appearance in the original image will confuse the network while training. Thus, we generate dynamically synthetic-like images for continual training. The overall scheme of SISSI framework is shown in Fig. 1.

3.3 Determining the Start of the Semi-supervised Phase

While training with mixed-batch training, we notice a two-stage learning phenomenon previously noted in classification and semantic segmentation: in an early learning phase, the network fits the clean annotations; then, the network start memorising the initial noisy annotations [6,7]. To find the optimal point that represents when the memorisation phase starts, we adopt a method, ADELE [6], that has been used in previous works in the context of semantic segmentation. In our work, we rely on the deceleration of the AP_{50} training curve of the model output and the initial noisy annotated dataset, CellLab, to decide when to stop trusting the initial noisy annotations and generate pseudo labels. See Fig. 7 in Appendix B for the AP_{50} training curve with the point representing when the memorisation phase starts.

3.4 Pseudo Label Generation

Pseudo label generation is a technique where a pre-trained neural network generates labels for unlabelled data or updates labels for noisy labeled data [16]. We

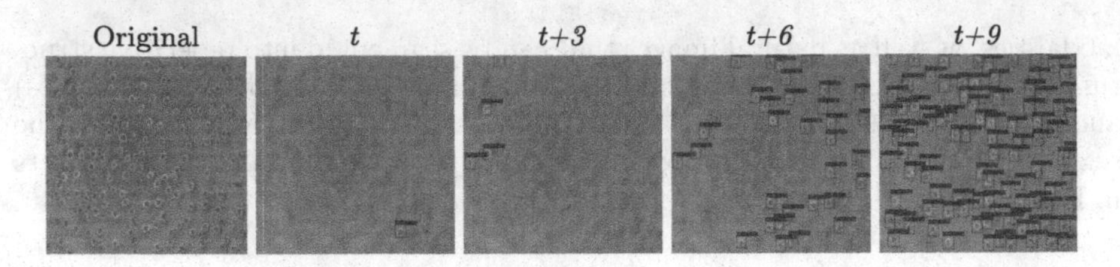

Fig. 3. Example of a synthetic-like image with weak blurring in training epochs (t).

generate pseudo labels to update the noisy annotations of the CellLab dataset during the semi-supervised phase. Self-training networks have the disadvantage of being unable to correct their own mistakes. Therefore biased and wrong labels can be amplified. To filter potential bounding boxes, we integrate two techniques, test-time augmentation (TTA) [17] and weighted boxes fusion (WBF) [13]. We average predictions generated with TTA while considering the confidence score of each bounding box in a WBF manner:

$$X_{1,2} = \frac{\sum_{i=1}^{T} C_i \cdot X_{1,2_i}}{\sum_{i=1}^{T} C_i}, \tag{1}$$

where T is the number for bounding boxes assigned to a single object in a cluster, $X_{1,2}$ (or $Y_{1,2}$) is the average start and end point on the x (or y) axis. This yields the average of the bounding box coordinates $X_{1,2_i}$ (or $Y_{1,2_i}$), weighted with the confidence score C_i for each bounding box.

3.5 Synthetic-Like Image Adaptation According to Pseudo Labels

Undetected cells in the pseudo labels would affect the further training negatively. When the network localises true objects that are not present in the pseudo-labels, the network is penalised for those objects that are true. To solve this problem, we propose to generate synthetic-like images dynamically according to the pseudo labels generated for the CellLab dataset, see Fig. 3. To remove unlabeled cells in the training image in order not to confuse the network, we clone all the detected cells of the pseudo label (source) onto a strongly/weakly Gaussian blurred image (target). To avoid discontinuities between the target and the source, we mix edge textures with the seamless cloning algorithm (mixing gradient) [9].

4 Experiments

4.1 Implementation Details

The backbone of our Faster R-CNN is a ResNet-50, pre-trained on the MS COCO dataset [5]. We set hyperparameters according to existing Fast/Faster R-CNN work [10]. We do not freeze any layer to allow the gradient to propagate through the early layers.

Algorithm 1. Pseudocode for iterative self-training with SISSI, prediction (p), target (t), bounding boxes (bbs).

Require: $CLimg, CLbbs_t, CPimg, CPbbs_t$ ▷ CellLab and Cellpose datasets
Require: $NN(img)$ ▷ Faster R-CNN
Require: $E \leftarrow this.self\_training\_epoch$
  for each pseudo_batch B in E **do** ▷ Pseudo label generation
    $CLbbs_p[n], scores[n] \leftarrow NN(TTA(CLimg \in B))$ ▷ Generate boxes with TTA
    $CLbbs_t \leftarrow (\sum_{n=1}^{N} scores[n] \cdot CLbbs_p[n]) / \sum_{n=1}^{N} scores[n]$ ▷ Filter bbs with WBF
    $update\_dataset(CLbbs_t)$ ▷ Update final pseudo label
  end for
  for each mixed_batch B in E **do** ▷ Training
    $CLimg\_crops[n] \leftarrow crop(CLimg, CLbbs)$ ▷ Synthetic image generation
    $CLimg\_blur \leftarrow blur(CLimg)$
    $CLimg\_synth \leftarrow seamless\_clone(CLimg\_blur, CLimg\_crops[n])$
    $CLbbs_p, CPbbs_p \leftarrow NN(CLimg\_synth, CPimg \in B)$ ▷ Prediction
    $CLCP\_loss \leftarrow loss([CLbbs_p, CPbbs_p], [CLbbs_t, CPbbs_t])$ ▷ Loss calculation
    $CLCP\_loss.backprop()$
  end for
  $E.next()$

We train the models using the Stochastic Gradient Descent (SGD) optimiser with a momentum of 0.9, weight decay of 0.0002, and learning rate of 0.001. We use a batch size of 8, with an equal number of images randomly chosen from the CellLab and Cellpose datasets, and resize the images to 512 × 512. We perform simple augmentations: channel shuffle, Gaussian blurring, horizontal flip, vertical flip, and shift-scale-rotate. For test-time augmentation used for label correction, we use a combination of scaling ([0.8, 0.9, 1, 1.1, 1.2]) and augmentations, vertical flipping, horizontal flipping, horizontal+vertical flipping, or no flipping. We end up with 20 versions of the same image. For background blurring in the synthetic-like image generation, we use Gaussian blurring with kernels of (21, 21) and kernels (12, 32), referred to as weak (W) and strong (S) background blurring respectively.

The datasets are used as follows. Mixed-batch training is applied in both the early supervised and semi-supervised learning phases, combining the CellLab and Cellpose training sets. With the start of self-training, labels and synthetic-like images for the CellLab dataset are updated in each following epoch. To perform validation for hyperparameter tuning, we use the Cellpose dataset since only five manually annotated images are available in the CellLab dataset, which all are used as a testset. For estimating the end of early learning, the weak training labels of CellLab are compared to the model output as proposed in ADELE. We calculate deceleration by the relative change in the derivative of the AP_{50} curve, and if it is above a certain threshold, 0.9, then label correction starts.

4.2 Evaluation Metrics and Results

In Table 1, we report three versions of AP, and AR over the CellLab testset. The metrics include the Pascal VOC metric (AP_{50}), as well as COCO evaluation metrics [5] (AP_{75}, and AP and AR averaged over different IoU thresholds). Bold numbers denote the best performance for each of the three cell experts' annotations.

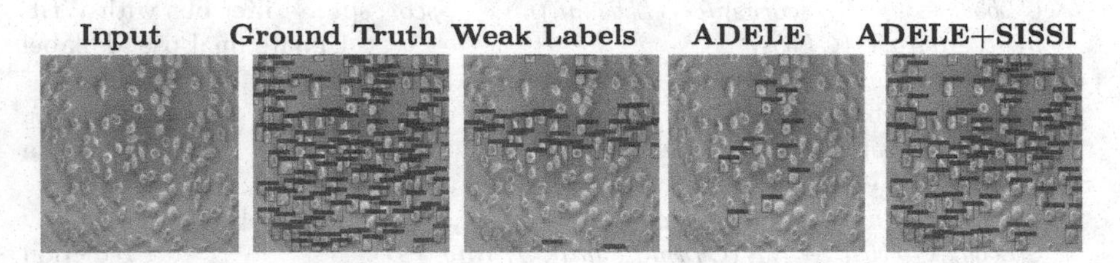

Fig. 4. Demonstration of improvement of results with our proposed method.

We present the detection performance of different experiments on the CellLab testset. The Baseline model is first trained in a supervised fashion with mixed-batch training of CellLab and Cellpose datasets. Aiming to correct and complete the labels, we perform early-stopping based on the validation loss of the source dataset, Cellpose, and apply self-training with test-time augmentation (TTA) and weighted boxes fusion (WBF) to iteratively update the pseudo labels. The following two experiments SISSI (W) and SISSI (S) additionally use synthetic-like image generation for the CellLab images, based on the pseudo labels generated with TTA and WBF. The weak (W) background blurring achieved better results than the baseline, while strong (S) background blurring has worse performance. Plain A, is an additional experiment similar to baseline setting but with an additional algorithm (ADELE) to find an optimal starting point for label correction. This shows an increase of about 10% across all readers in the AR metrics, compared to the baseline model, while the AP is lower.

Two versions of the final pipeline show the best results. It combines all steps (1) supervised learning with mixed-batch training of CellLab and Cellpose till a memorisation point is reached, (2) iteratively applying pseudo-label generation for the CellLab dataset with test-time augmentation and weighted boxes fusion, (3) generation of synthetic-like images according to the pseudo labels, and (4) training the network for an Epoch with mixed-batch training of the Cellpose dataset and the pseudo labels and synthetic-like images of the CellLab dataset. Pseudo code for the loop of 2–4 can be seen in Algorithm 1. Incorporating ADELE with our label correction and synthetic-like image generation method with strong blurring increases the AP by at least 15% and AR by at least 20% compared to the baseline across all cell experts. On Fig. 4, we show an example where SISSI successfully improves the detection results of Plain A (ADELE) experiment. Examples of a training image with its pseudo labels for different epochs (t) and experiments can be seen in Fig. 8 in Appendix B.

5 Discussion

5.1 Findings

The experiments made clear that both the start of label correction and the amount of background information appearing in images during training impact the results. When starting label correction too early, during early learning, the network is not confident enough to detect all objects in the image; thus, correcting initially noisy annotations at this stage results in a high rate of missing targets. Training a network on images with high missing targets without SISSI (Plain A) increases the uncertainty of the network compared to label correction in a later memorization phase with fewer missing targets, the baseline.

Table 1. Results of Cell Detection on the CellLab testset.

	Annotator 1				Annotator 2				Annotator 3			
Pipeline	AP_{50}	AP_{75}	AP	AR	AP_{50}	AP_{75}	AP	AR	AP_{50}	AP_{75}	AP	AR
Baseline	45.6	15.7	21.3	32.7	44.3	16.1	20.2	31.0	**58.6**	28.1	29.7	41.8
SISSI (W)	52.8	23.2	25.9	37.7	49.5	21.3	24.4	34.6	58.5	29.0	29.7	42.0
SISSI (S)	40.3	8.3	16.2	32.8	38.4	7.2	15.2	31.8	46.6	9.7	19.2	37.8
Plain A	38.7	18.9	19.2	43.9	35.8	18.6	18.7	43.1	41.2	23.8	22.9	50.6
A+SISSI (W)	43.1	37.1	36.0	**60.3**	45.1	38.5	37.4	**58.8**	47.6	42.8	41.4	**66.9**
A+SISSI (S)	**54.9**	**49.0**	**43.2**	57.6	**51.2**	**45.1**	**39.7**	54.3	58.5	**55.5**	**47.9**	64.9

In the basic SISSI approach, where label correction is started on a model chosen based on the validation loss of the external Cellpose dataset, weak background blurring worked better than a strongly blurred background. We believe this phenomenon appears because the neural network has learned more contextual information in the memorisation stage and requires the background information.

On the other hand, starting label correction when the early learning phase ends, according to ADELE, strong blurring shows better results than weak blurring. The information about the background is less important. This can be an advantage in synthetic-like image generation because figuring out how to preserve contextual information seems less critical.

5.2 Limitations

The success of SISSI may be dependent on the stopping criteria and the training phase, early learning/memorisation phase. When the annotations in the image are too noisy, the network may not encompass the early learning phase as in previous works, ADELE. It may be unable to learn the task to produce new pseudo labels for further training. The effect of blurring during different training phases needs more empirical research for verification. SISSI is a simple approach that works with only one class of interest to detect. Blurring with multi-object needs further modification in future works. We use SISSI in these experiments with Faster R-CNN, which is more robust and friendly for the missing label scenario than other detection networks.

5.3 Conclusion

This paper presents a method to train object detection models with noisy and missing annotations with semi-supervised learning by proposing a novel technique. We use dynamically generated synthetic-like images using seamless cloning for further training the network after pseudo-label generation. We utilize a domain adaptation technique, Balanced Gradient Contribution, to generate stable gradient directions and mitigate the so noisy annotation problem for our semi-supervised training. Finally, we evaluate our method for the cell detection task with various training procedures and show its improvement over the usual semi-supervised approach. Our method, SISSI, can be added on top of any detection network, and it also helps other methods like ADELE to be leveraged for object detection. In the future, we will adapt our method to work with multi-object detection and explore SISSI with different detection networks. Moreover, we will explore our method for different medical detection tasks and integrate our network to help cell experts with the grading task.

Acknowledgements. We would like to acknowledge Oesterreichisches Forschungsinstitut für Chemie und Technik (OFI) for CellLab images and test set annotation.

References

1. Anderson, J.M.: Future challenges in the in vitro and in vivo evaluation of biomaterial biocompatibility. Regen. Biomater. **3**(2), 73–77 (2016). https://doi.org/10.1093/rb/rbw001
2. Gao, J., Wang, J., Dai, S., Li, L.J., Nevatia, R.: NOTE-RCNN: noise tolerant ensemble RCNN for semi-supervised object detection. In: Proceedings of the IEEE/CVF International Conference on Computer Vision (ICCV), October 2019
3. ISO: ISO 10993-5: 2009-biological evaluation of medical devices-part 5: tests for in vitro cytotoxicity (2009)
4. Li, H., Pan, X., Yan, K., Tang, F., Zheng, W.S.: SIOD: single instance annotated per category per image for object detection. In: Proceedings of the IEEE/CVF Conference on Computer Vision and Pattern Recognition (CVPR), pp. 14197–14206 (2022)
5. Lin, T.-Y., et al.: Microsoft COCO: common objects in context. In: Fleet, D., Pajdla, T., Schiele, B., Tuytelaars, T. (eds.) ECCV 2014. LNCS, vol. 8693, pp. 740–755. Springer, Cham (2014). https://doi.org/10.1007/978-3-319-10602-1_48
6. Liu, S., Liu, K., Zhu, W., Shen, Y., Fernandez-Granda, C.: Adaptive early-learning correction for segmentation from noisy annotations. In: Proceedings of the IEEE/CVF Conference on Computer Vision and Pattern Recognition, pp. 2606–2616 (2022)

7. Liu, S., Niles-Weed, J., Razavian, N., Fernandez-Granda, C.: Early-learning regularization prevents memorization of noisy labels. In: Advances in Neural Information Processing Systems, vol. 33 (2020)
8. Lyu, F., Yang, B., Ma, A.J., Yuen, P.C.: A segmentation-assisted model for universal lesion detection with partial labels. In: de Bruijne, M., et al. (eds.) MICCAI 2021. LNCS, vol. 12905, pp. 117–127. Springer, Cham (2021). https://doi.org/10.1007/978-3-030-87240-3_12
9. Pérez, P., Gangnet, M., Blake, A.: Poisson image editing. ACM Trans. Graph. **22**(3), 313–318 (2003). https://doi.org/10.1145/882262.882269
10. Ren, S., He, K., Girshick, R., Sun, J.: Faster R-CNN: towards real-time object detection with region proposal networks. In: Advances in Neural Information Processing Systems, vol. 28 (2015)
11. Ros, G., Stent, S., Fernández Alcantarilla, P., Watanabe, T.: Training constrained deconvolutional networks for road scene semantic segmentation. CoRR (2016)
12. Siegismund, D., Tolkachev, V., Heyse, S., Sick, B., Duerr, O., Steigele, S.: Developing deep learning applications for life science and pharma industry. Drug Res. **68**(06), 305–310 (2018)
13. Solovyev, R., Wang, W., Gabruseva, T.: Weighted boxes fusion: ensembling boxes from different object detection models. Image Vis. Comput. **107**, 104117 (2021)
14. Stringer, C., Pachitariu, M.: Cellpose 2.0: how to train your own model. bioRxiv (2022). https://doi.org/10.1101/2022.04.01.486764, https://www.biorxiv.org/content/early/2022/04/05/2022.04.01.486764
15. Tajbakhsh, N., Jeyaseelan, L., Li, Q., Chiang, J.N., Wu, Z., Ding, X.: Embracing imperfect datasets: a review of deep learning solutions for medical image segmentation. Med. Image Anal. **63**, 101693 (2020). https://doi.org/10.1016/j.media.2020.101693, https://www.sciencedirect.com/science/article/pii/S136184152030058X
16. Triguero, I., García, S., Herrera, F.: Self-labeled techniques for semi-supervised learning: taxonomy, software and empirical study. Knowl. Inf. Syst. **42**(2), 245–284 (2013). https://doi.org/10.1007/s10115-013-0706-y
17. Wang, G., Li, W., Aertsen, M., Deprest, J., Ourselin, S., Vercauteren, T.: Aleatoric uncertainty estimation with test-time augmentation for medical image segmentation with convolutional neural networks. Neurocomputing **338**, 34–45 (2019)
18. Wang, Y., et al.: Semi-supervised semantic segmentation using unreliable pseudo-labels. In: Proceedings of the IEEE/CVF Conference on Computer Vision and Pattern Recognition (CVPR), pp. 4248–4257, June 2022
19. Xiong, H., Liu, S., Sharan, R.V., Coiera, E., Berkovsky, S.: Weak label based Bayesian U-Net for optic disc segmentation in fundus images. Artif. Intell. Med. **126**, 102261 (2022)
20. Zhang, H., Chen, F., Shen, Z., Hao, Q., Zhu, C., Savvides, M.: Solving missing-annotation object detection with background recalibration loss. In: ICASSP 2020 - 2020 IEEE International Conference on Acoustics, Speech and Signal Processing (ICASSP), pp. 1888–1892 (2020)

Task-Agnostic Continual Hippocampus Segmentation for Smooth Population Shifts

Camila González[1(✉)], Amin Ranem[1], Ahmed Othman[2], and Anirban Mukhopadhyay[1]

[1] Darmstadt University of Technology, Karolinenplatz 5, 64289 Darmstadt, Germany
camila.gonzalez@gris.tu-darmstadt.de

[2] University Medical Center Mainz, Langenbeckstraße 1, 55131 Mainz, Germany

Abstract. Most continual learning methods are validated in settings where task boundaries are clearly defined and task identity information is available during training and testing. We explore how such methods perform in a task-agnostic setting that more closely resembles dynamic clinical environments with gradual population shifts. We propose ODEx, a holistic solution that combines out-of-distribution detection with continual learning techniques. Validation on two scenarios of hippocampus segmentation shows that our proposed method reliably maintains performance on earlier tasks without losing plasticity.

Keywords: Continual learning · Lifelong learning · Distribution shift

1 Introduction

Deep learning methods are mostly validated in stationary environments where the train and test data have been carefully homogenized to preserve the i.i.d. assumption. This does not reflect the reality of clinical deployment, where acquisition conditions and disease patterns evolve over time. *Continual learning* (CL) paradigms are being explored by medical imaging researchers [19,22,27] and regulatory bodies [29] as evaluation settings that are better suited for AI in healthcare. Continual methods deal with temporal restrictions on data availability by sequentially accumulating knowledge over a stream of *tasks*, each containing data from a different distribution, without revisiting previous stages.

Yet most CL approaches are validated in settings with *rigid task boundaries* and *known task labels*, which is far from how real dynamic environments behave [7]. When deviating from this simplistic problem formulation, they perform worse

Supported by the Bundesministerium für Gesundheit (BMG) with grant [ZMVI1-2520DAT03A].

Supplementary Information The online version contains supplementary material available at https://doi.org/10.1007/978-3-031-16852-9_11.

K. Kamnitsas et al. (Eds.): DART 2022, LNCS 13542, pp. 108–118, 2022.
https://doi.org/10.1007/978-3-031-16852-9_11

than simple baselines [23]. Previous research has established desirable properties for CL methods, illustrated in Fig. 1. These include no reliance on either (1) assumptions on task boundaries during training or (2) access to task identity labels, i.e. the method should be *task-agnostic* [10]. In addition, the model should (3) preserve previous knowledge while (4) maintaining sufficient plasticity to learn new tasks and (5) not require additional computational resources during training [7,10]. The last three objectives are often deemed to be orthogonal, i.e. most approaches either *catastrophically forget* previous knowledge (too plastic), cannot learn new tasks (too rigid) or the training time and resource requirements grow linearly with the number of tasks.

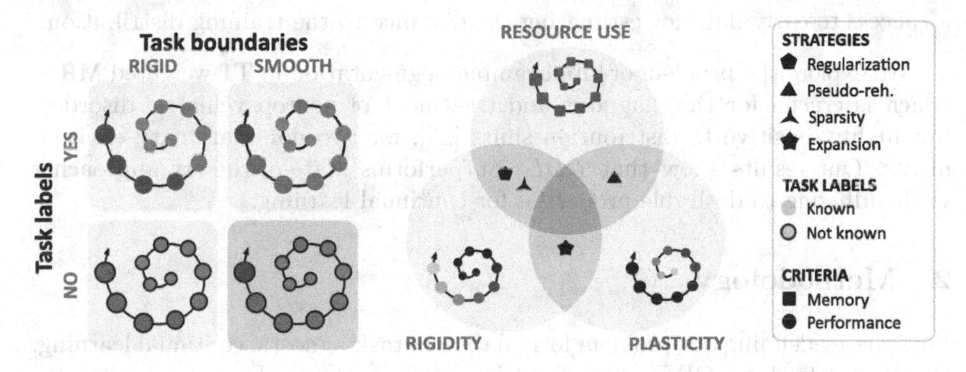

Fig. 1. Desiderata for continual learning [7,10]. Left: methods should not rely on rigid boundaries or task labels. Right: trade-off between plasticity, rigidity and resource use.

Methods for task-agnostic continual learning are overwhelmingly *rehearsal-based* [1,2,12,21,27], i.e. store a subset of past images or features in a memory buffer, which is not admissible in many diagnostic settings due to patient privacy considerations. *Active learning* methods also exist which rely on expert interaction [22].

Other approaches train generative models to identify distribution shifts [24] or only update the shortest sub-path of the network that allows a correct classification [6], but such solutions are computationally expensive and are therefore only evaluated in low-resolution classification settings. The field of continual learning for medical segmentation is still under-studied. Most research follows regularization-based strategies that calculate the importance of parameters and penalize their deviation [19,30]. Approaches have also been proposed for active learning [31], others allow the storage of previous samples [21,28]. Some methods leverage feature disentanglement to alleviate forgetting [16,18] or maintain task-dependent batch normalization layers [13]. To our knowledge, no method has been previously introduced for semantic segmentation that is task-agnostic and does not make use of a rehearsal component.

We propose **ODEx**, an expansion-based approach that (1) does not revisit previous stages, (2) is well-suited to a wide array of use cases, including semantic segmentation and (3) is task-agnostic, i.e. requires neither task boundaries nor task labels during training or inference. *ODEx* uses continual out-of-distribution (OOD) detection to signal when to *expand* the model and select the best parameters during inference. Although we maintain multiple parameter states in persistent memory, each occupies less than 0.2 GB and the continual OOD detection mechanism ensures that this number remains low. Unlike other methods, *ODEx* requires the same GPU memory and training time as regular sequential learning. Our contributions include:

1. proposing a task-agnostic continual learning solution suitable for a wide array of deep learning architectures, and
2. introducing a continual OOD detection mechanism that does not require access to early data for estimating the distance to the training distribution.

We explore the problem of hippocampus segmentation in T1-weighted MRIs, which is crucial for the diagnosis and treatment of neuropsychiatric disorders but highly sensitive to distribution shifts [25], for two non-stationary environments. Our results show that *ODEx* outperforms state-of-the-art approaches while adhering to desirable properties for continual learning.

2 Methodology

We start by defining our problem formulation of task-agnostic continual learning. We then introduce *ODEx*, visualized in Fig. 2 (bottom). During training, we accumulate the mean and covariance of batch normalization layers and detect domain shifts with the Mahalanobis distance. When a domain shift occurs, a new model is initialized with the most appropriate parameters and added to the model pool. During inference, we extract predictions with the best model state.

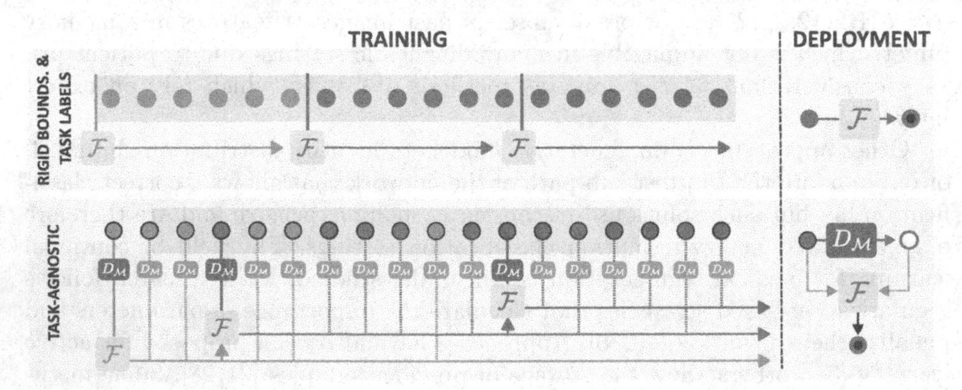

Fig. 2. Top: continual setting with rigid boundaries and task labels. Expansion methods create new parameters at each task boundary. Bottom: the task-agnostic *ODEx* method initializes a new set of parameters when a domain shift is detected.

Task-Agnostic Continual Learning: In continual learning settings, model $\mathcal{F}_\theta : x \rightarrow \hat{y}$ is trained with data samples from an array of N_t different *tasks* or data distributions $\{\mathcal{T}_i ... \mathcal{T}_{N_t}\}$, each found at the i_{th} *stage* t_i. The model should be deployable after finishing the first stage, and evolve over time. For segmentation, each instance has the form (x, y, j), where x is an image and y the segmentation mask. Additionally, j denotes the *task label*, i.e. that $(x, y) \sim \mathcal{T}_j$. The goal is to find parameters θ that minimize the loss $\mathcal{L}$ over all seen tasks $\{\mathcal{T}_i\}_{i \leq N_t}$ (Eq. 1).

$$\arg\min_{\theta} \sum_{j=1}^{N_t} \mathbb{E}_{(x,y) \sim \mathcal{T}_j} \left[\mathcal{L}\left(\mathcal{F}_{\theta}(x), y\right)\right] \tag{1}$$

The objective cannot be optimized directly, as at any training stage t_j only data from $\mathcal{T}_j$ is available. The main challenge consists of ensuring enough *rigidity* during training to obtain good performance on $(x, y) \sim \{\mathcal{T}_i\}_{i<j}$ and enough *plasticity* to learn from present and future data $(x, y) \sim \{\mathcal{T}_i\}_{i \geq j}$.

Expansion-based methods approach this by keeping *task-dependent* parameters $\{\theta_1 ... \theta_{N_t}\}$, which in their simplest form comprise the entire model, and perform inference on (x, y, j) with the respective $\mathcal{F}_{\theta_j}$ (see Fig. 2, above). In *task-agnostic scenarios*, task labels j are unknown and may not even be clearly defined. The goal is to learn a set of parameters $\Theta = \{\theta_1 ... \theta_{|\Theta|}\}$ and an inference function $\mathcal{J} : x \rightarrow \theta$ that selects the best parameters during testing (Eq. 2). In the absence of rigid task boundaries, the size of the model pool $|\Theta|$ is unknown. Task-agnostic settings thus signify three additional challenges: (1) detecting when domain shifts occur, (2) keeping $|\Theta|$ low and (3) choosing the best parameters during testing. In the following, we outline how we approach these.

$$\arg\min_{\Theta} \sum_{j=1}^{N_t} \mathbb{E}_{(x,y) \sim \mathcal{T}_j} \left[\mathcal{L}\left(\mathcal{F}_{\mathcal{J}(x)}(x), y\right)\right] \tag{2}$$

Detecting Domain Shifts: During training, we extract features z from the first set of *Batch Normalization* layers BN_1. These normalize inputs and thus contain domain-pertinent information which has been found to play a key role in detecting interference during sequential learning [13]. We estimate a multivariate Gaussian $\mathcal{N}_i(\mu_i, \Sigma_i)$ at the end of training stage t_i as:

$$z_k \leftarrow BN_1(x_k); \mu_i \leftarrow \frac{1}{N} \sum_{k=1}^{N} z_k; \Sigma_i \leftarrow \frac{1}{N} \sum_{k=1}^{N} (z_k - \mu_i)(z_k - \mu_i)^T \tag{3}$$

Inspired by previous research on OOD detection for semantic segmentation [9], we detect data shifts by calculating the *Mahalanobis distance* $D_{\mathcal{M}}(z; \mu, \Sigma)$ to the training distribution. In contrast to other methods for assessing similarity, such as the *Gram distance* popular in rehearsal-based continual learning [21,22], the Mahalanobis distance requires storing only μ and Σ.

As we cannot revisit data from previous stages, we cannot estimate $\mathcal{N}$ with all data used to train the model. In a situation with slowly shifting data distributions, if we were to only consider the μ and Σ of the last training batch, then we may never detect a sufficiently large distance signaling the need to expand the model pool. We therefore store μ_i and Σ_i at the end of each training stage t_i

and add this to the *history* $\mathcal{B}_i$ of the model which contains information from all pertinent training stages. At stage t_{i+1}, parameters $\hat{\theta}$ are selected that minimize the summed distance of the present training data to the history of $\hat{\theta}$ (Eq. 4).

$$D_{\mathcal{M}}(z;i) : \min_{\theta_j \in \Theta_i} \sum_{(\mu_j, \Sigma_j) \in \mathcal{B}_j} D_{\mathcal{M}_j}(z;\mu_j,\Sigma_j) \tag{4}$$

Managing the Model Pool: When data arrives for a new stage t_i, the distance $D_{\mathcal{M}}(z;i)$ is calculated and the best model $\hat{\theta}$ is selected. If $D_{\mathcal{M}}(z;i) < \xi$ (case 1), then $\hat{\theta}$ is updated with the current data. Afterwards, μ_i and Σ_i are calculated and added to the model history $\hat{\mathcal{B}}$. If instead $D_{\mathcal{M}}(z;i) \geq \xi$ (case 2), a domain shift is detected and a new model θ_i is initialized with the parameters of $\hat{\theta}$. After a domain shift, the size of the model pool $|\Theta|$ grows by 1. The history of the new model $\mathcal{B}_i$ is initialized with $\hat{\mathcal{B}}$, so the history of each model contains information pertaining to all data distributions used to train it. Following previous research [9] we normalize the distances between the minimum and doubled maximum in-distribution values, and set $\xi = 2\mu$.

Continuing to train older models instead of initializing a new one for each stage has two advantages: (1) the model pool does not grow linearly with the length of the data stream, which would be prohibiting for deployment over long time periods and (2) models can benefit from further training when the data distributions are compatible, potentially allowing positive backwards transfer.

Performing Inference: Inference proceeds as illustrated in Fig. 2 (right). For each image, the summed Mahalanobis distance of the test image to each set of parameters $\theta \in \Theta$ is calculated. Again, the best model $\hat{\theta}$ is selected and, in this case, directly used to extract a segmentation mask $\mathcal{F}_{\hat{\theta}}(x) = \hat{y}$.

3 Experimental Setup

We briefly outline how we build our data base of tasks with smooth distribution shifts from publicly available datasets and report relevant aspects of our experimental setup. For further implementation details, we refer the reader to the supplementary material and our code found under https://github.com/MECLabTUDA/Lifelong-nnUNet.

Data: We look at two different scenarios of data streams with slowly shifting distributions for segmentation of the entire hippocampus (head, body and tail) in T1-weighted MRIs. The first is constructed from three public datasets: *HarP* [3] contains 135 healthy and Alzheimer's disease patients, *Dryad* [15] has 25 healthy adult subjects and *Decathlon* [26] contains 130 healthy and schizophrenia patients. We slowly shift the distribution of cases from each source as illustrated in Appendix A. We refer to this scenario as **shifting source**. For the second scenario, henceforth referred to as **transformed**, we slowly modify the *Decathlon* data using the *TorchIO* library [20]. We apply intensity rescaling up to a contrast stretching of (0.1, 0.9) and affine transformations of up to a (0.8, 1.2) scaling range, 15 °C rotation and 5 mm translation.

Table 1. Performance of the joint training upper bound (first row), sequential learning and six continual learning strategies on the two hippocampus segmentation scenarios.

Method	Shifting source			Transformed		
	Dice ↑	BWT ↑	FWT ↑	Dice ↑	BWT ↑	FWT ↑
Joint	.89 ± .01			.90 ± .01		
Seq.	.57 ± .32	−.19 ± .12	**.14** ± .09	.87 ± .03	−.02 ± .02	.09 ± .05
EWC	.78 ± .08	**.02** ± .03	.08 ± .08	.79 ± .10	**.01** ± .01	.04 ± .02
MiB	.67 ± .24	−.10 ± .07	**.14** ± .10	.87 ± .04	−.02 ± .02	.07 ± .04
RW	.61 ± .28	−.15 ± .10	**.14** ± .10	.87 ± .03	−.03 ± .03	.09 ± .05
PLOP	.57 ± .32	−.22 ± .14	.13 ± .09	.86 ± .02	−.02 ± .02	**.10** ± .06
LwF	.51 ± .35	−.23 ± .13	.10 ± .07	.86 ± .04	−.04 ± .04	**.10** ± .06
ODEx (ours)	**.87** ± .04	−.03 ± .02	**.14** ± .09	**.89** ± .01	−.01 ± .01	.09 ± .05

Network Architecture and Training: We use a full-resolution *nnUNet* [11] model for all experiments, with the architecture and training settings selected for the first training stage of each data stream. We perform 200 epochs for each stage, with a loss of *Dice* and *Binary Cross Entropy* weighted equally. All experiments were carried out on a *Nvidia Tesla T4* GPU (16 GB).

Metrics: We report the average Dice on test data from all tasks $\{T_i\}_{i \leq N_t}$ as well as backwards (BWT) and forwards (FWT) transferability [7,10]. BWT is the *inverse forgetting* and displays to what extent the performance on test samples $(x, y) \sim T_i$ deteriorates with further training in stages $\{t_i\}_{i>N_t}$. FWT instead measures what impact training on each stage $\{t_i\}_{i \leq N_t}$ has on test data $(x, y) \sim T_i$. Methods that prevent forgetting show high, realistically close to 0, BWT. FWT is high if enough plasticity is maintained to acquire new knowledge. For both metrics, we report the average over test data from all tasks.

Baselines: In Sect. 4.1, we compare our approach against sequential training and five popular continual learning approaches: Elastic Weight Consolidation (EWC) [14], Modelling the Background (MiB) [4], Riemannian Walk (RW) [5], PLOP [8] and Learning without Forgetting (LwF) [17]. We also report the upper bound of joint training. In most cases, we use the hyperparameters suggested in the corresponding publications or code bases (for more details see Appendix B). For MiB, we reduce the *lkd* to prevent loss explosion. In Sect. 4.2 we perform an ablation study and compare the use of the Mahalanobis distance to other methods proposed within task-agnostic learning, namely using the Gram matrix [21] and detecting domain shifts through a fall in training performance [6].

4 Results

We first compare *ODEx* to state-of-the-art continual learning approaches in Sect. 4.1. Afterwards, we take a closer look at the cumulative Mahalanobis distance for identifying domain shifts and selecting the best parameters (Sect. 4.2).

4.1 Continual Learning Performance

We compare our proposed approach *ODEx* to five continual learning methods in Table 1. The first row shows the upper bound of training a model statically with all training data. *Sequential* results show the deterioration of the performance in earlier tasks as training is carried out, and the following rows display how five continual learning strategies alleviate this. From these, only *EWC* maintains performance on earlier tasks, but at the cost of losing model plasticity and being unable to acquire new knowledge. *ODEx* instead reaches a high FWT showing effective learning on later tasks while still performing well on data from the first training stages. This behavior is further illustrated in Fig. 3 (left), where the per-task performance is plotted for *EWC*, which successfully retains old knowledge, *MiB*, which reaches a high Dice on later tasks, and *ODEx* that performs well on data from all stages. This is particularly clear for the more difficult *shifting source* case, but a Wilcoxon one-sided signed-rank test affirms that *ODEx* significantly outperforms all other approaches in terms of Dice score for both scenarios.

As for resource utilization, *ODEx* requires no more GPU memory than sequential training, as we update one model at a time. The estimation of Σ and the calculation of $D_{\mathcal{M}}$ can be carried out in the CPU given the low resolution of z. Figure 3 (right) shows that *ODEx* takes only marginally longer than training without any method for forgetting prevention. Though several models are stored (two for *shifting source* and four for *transformed*, see Table 2) each weights less than 200 MB, being far from a limiting factor in practice.

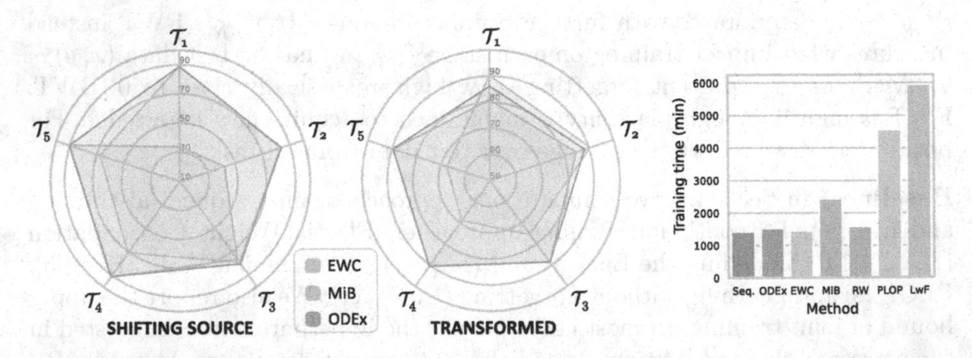

Fig. 3. Left: Per-task Dice. EWC and MiB are at opposite ends of the plasticity/rigidity spectrum, whereas ODEx allows for further training without compromising performance on previous tasks. Right: training times for the shifting source scenario.

Figure 4 qualitatively shows in the upper row the sequential deterioration of the segmentation for a test subject $(x, y) \sim \mathcal{T}_1$. The lower row displays the segmentation masks produced by each continual learning method. Though the head is mostly segmented well by several methods, only *EWC* and *ODEx* properly segment the body and tail and maintain the integrity of the shape.

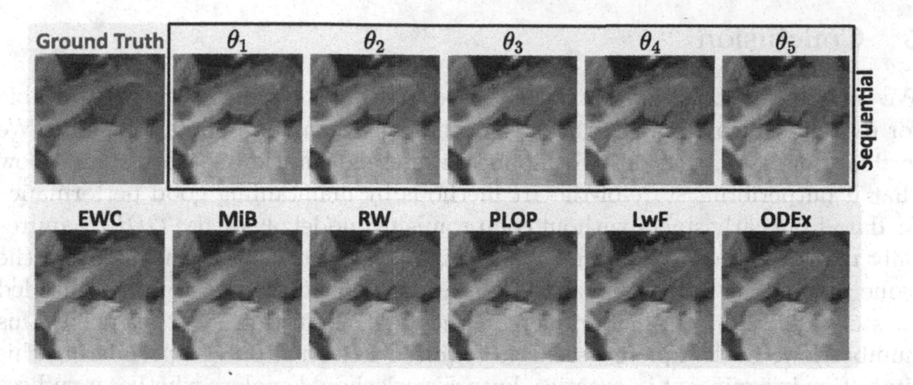

Fig. 4. Crops with overlayed segmentations for axial slice 25 of a subject from $\mathcal{T}_1$ (shifting source). Top: ground truth (blue) and performance deterioration with regular SGD. Bottom: six continual learning methods, after finishing training on last stage. (Color figure online)

4.2 Ablation Study

In Table 2 we compare our strategy for detecting when to grow the model pool to previous work in the field of task-agnostic learning. The performance of all methods is very similar for the easier *transformed* scenario, but we see clear differences in *shifting source*. We first explore two versions of *ODEx* that use our proposed strategy for selecting the best model but detect domain shifts in a different fashion. *ODEx* $-\infty$ ξ creates a new model for every stage. The lower Dice suggests that the models suffer from the lack of training data, and $|\Theta|$ grows linearly with the number of training stages. *DiceEx* initializes a new model when the training Dice falls more than 10%, which results in higher forgetting. *ODEx* $\neg\mathcal{B}$ shows the situation where we do not keep a history for the training distributions of previous stages and only calculate the distance to the last stage. For this version, no new model is initialized for *shifting source* and the single available model significantly forgets previous knowledge. Finally, we test the use of the Gram distance instead of Mahalanobis for both training and testing, and find that it does not properly detect distribution shifts for *shifting source.*

Table 2. Performance of different strategies for detecting domain boundaries and/or selecting a model state during inference.

	Shifting source				Transformed			
Method	Dice ↑	BWT ↑	FWT ↑	$\|\Theta\|$↓	Dice ↑	BWT ↑	FWT ↑	$\|\Theta\|$↓
ODEx (ours)	**.87** ± .04	−.03 ± .02	**.14** ± .09	2	.89 ± .01	−.01 ± .01	**.09** ± .05	4
ODEx $-\infty$ ξ	.83 ± .04	**.00** ± .00	.11 ± .09	5	.89 ± .01	**.00** ± .00	**.09** ± .05	5
DiceEx [6]	.84 ± .08	−.07 ± .03	**.14** ± .10	2	.89 ± .02	−.01 ± .01	**.09** ± .05	**2**
ODEx $\neg\mathcal{B}$ [9]	.57 ± .32	−.19 ± .12	**.14** ± .09	**1**	.89 ± .01	**.00** ± .00	**.09** ± .05	3
Gram [21]	.57 ± .32	−.19 ± .12	**.14** ± .09	**1**	**.90** ± .01	**.00** ± .00	**.09** ± .05	3

5 Conclusion

We introduce *ODEx*, an expansion-based continual learning strategy suitable for real clinical environments with smooth acquisition and population shifts. We evaluate our approach on two hippocampus segmentation scenarios and show that it outperforms state-of-the-art methods by maintaining good performance on data from early stages without compromising model plasticity. *ODEx* requires only marginally higher training times than regular sequential learning, and the same amount of GPU memory. While additional persistent storage is needed to store different sets of parameters, the OOD detection strategy keeps this number low. Each explored scenario required less than 0.8 GB, rendering this limitation insignificant in practice. Future work should explore whether it suffices to maintain only a subset of domain-specific parameters, such as the last decoder blocks or batch normalization layers. By releasing our code and models, we hope to boost continual learning research in task-agnostic medical settings.

References

1. Aljundi, R., et al.: Online continual learning with maximal interfered retrieval. In: NeurIPS, vol. 32 (2019)
2. Aljundi, R., Lin, M., Goujaud, B., Bengio, Y.: Gradient based sample selection for online continual learning. In: NeurIPS, vol. 32 (2019)
3. Boccardi, M., et al.: Training labels for hippocampal segmentation based on the EADC-ADNI harmonized hippocampal protocol. Alzheimer's Dementia **11**(2), 175–183 (2015)
4. Cermelli, F., Mancini, M., Bulo, S.R., Ricci, E., Caputo, B.: Modeling the background for incremental learning in semantic segmentation. In: CVPR, pp. 9233–9242 (2020)
5. Chaudhry, A., Dokania, P.K., Ajanthan, T., Torr, P.H.S.: Riemannian walk for incremental learning: understanding forgetting and intransigence. In: Ferrari, V., Hebert, M., Sminchisescu, C., Weiss, Y. (eds.) ECCV 2018. LNCS, vol. 11215, pp. 556–572. Springer, Cham (2018). https://doi.org/10.1007/978-3-030-01252-6_33
6. Chen, H.J., Cheng, A.C., Juan, D.C., Wei, W., Sun, M.: Mitigating forgetting in online continual learning via instance-aware parameterization. In: NeurIPS, vol. 33, pp. 17466–17477 (2020)
7. Delange, M., et al.: A continual learning survey: defying forgetting in classification tasks. IEEE Trans. Pattern Anal. Mach. Intell. **44**, 3366–3385 (2021)
8. Douillard, A., Chen, Y., Dapogny, A., Cord, M.: PLOP: learning without forgetting for continual semantic segmentation. In: CVPR, pp. 4040–4050 (2021)
9. Gonzalez, C., Gotkowski, K., Bucher, A., Fischbach, R., Kaltenborn, I., Mukhopadhyay, A.: Detecting when pre-trained nnU-net models fail silently for Covid-19 lung lesion segmentation. In: de Bruijne, M., et al. (eds.) MICCAI 2021. LNCS, vol. 12907, pp. 304–314. Springer, Cham (2021). https://doi.org/10.1007/978-3-030-87234-2_29
10. Hadsell, R., Rao, D., Rusu, A.A., Pascanu, R.: Embracing change: continual learning in deep neural networks. Trends Cogn. Sci. **24**(12), 1028–1040 (2020)
11. Isensee, F., Jaeger, P.F., Kohl, S.A., Petersen, J., Maier-Hein, K.H.: nnU-Net: a self-configuring method for deep learning-based biomedical image segmentation. Nat. Methods **18**(2), 203–211 (2021)

12. Jin, X., Sadhu, A., Du, J., Ren, X.: Gradient-based editing of memory examples for online task-free continual learning. In: NeurIPS, vol. 34 (2021)
13. Karani, N., Chaitanya, K., Baumgartner, C., Konukoglu, E.: A lifelong learning approach to brain MR segmentation across scanners and protocols. In: Frangi, A.F., Schnabel, J.A., Davatzikos, C., Alberola-López, C., Fichtinger, G. (eds.) MICCAI 2018. LNCS, vol. 11070, pp. 476–484. Springer, Cham (2018). https://doi.org/10.1007/978-3-030-00928-1_54
14. Kirkpatrick, J., et al.: Overcoming catastrophic forgetting in neural networks. Proc. Natl. Acad. Sci. **114**(13), 3521–3526 (2017)
15. Kulaga-Yoskovitz, J., et al.: Multi-contrast submillimetric 3 tesla hippocampal subfield segmentation protocol and dataset. Sci. Data **2**(1), 1–9 (2015)
16. Lao, Q., Jiang, X., Havaei, M., Bengio, Y.: Continuous domain adaptation with variational domain-agnostic feature replay. arXiv preprint arXiv:2003.04382 (2020)
17. Li, Z., Hoiem, D.: Learning without forgetting. IEEE Trans. Pattern Anal. Mach. Intell. **40**(12), 2935–2947 (2017)
18. Memmel, M., Gonzalez, C., Mukhopadhyay, A.: Adversarial continual learning for multi-domain hippocampal segmentation. In: Albarqouni, S., et al. (eds.) DART/FAIR -2021. LNCS, vol. 12968, pp. 35–45. Springer, Cham (2021). https://doi.org/10.1007/978-3-030-87722-4_4
19. Özgün, S., Rickmann, A.-M., Roy, A.G., Wachinger, C.: Importance driven continual learning for segmentation across domains. In: Liu, M., Yan, P., Lian, C., Cao, X. (eds.) MLMI 2020. LNCS, vol. 12436, pp. 423–433. Springer, Cham (2020). https://doi.org/10.1007/978-3-030-59861-7_43
20. Pérez-García, F., Sparks, R., Ourselin, S.: TorchIO: a python library for efficient loading, preprocessing, augmentation and patch-based sampling of medical images in deep learning. Comput. Methods Programs Biomed. **208**, 106236 (2021). https://doi.org/10.1016/j.cmpb.2021.106236. https://www.sciencedirect.com/science/article/pii/S0169260721003102
21. Perkonigg, M., et al.: Dynamic memory to alleviate catastrophic forgetting in continual learning with medical imaging. Nat. Commun. **12**(1), 1–12 (2021)
22. Perkonigg, M., Hofmanninger, J., Langs, G.: Continual active learning for efficient adaptation of machine learning models to changing image acquisition. In: Feragen, A., Sommer, S., Schnabel, J., Nielsen, M. (eds.) IPMI 2021. LNCS, vol. 12729, pp. 649–660. Springer, Cham (2021). https://doi.org/10.1007/978-3-030-78191-0_50
23. Prabhu, A., Torr, P.H.S., Dokania, P.K.: GDumb: a simple approach that questions our progress in continual learning. In: Vedaldi, A., Bischof, H., Brox, T., Frahm, J.-M. (eds.) ECCV 2020. LNCS, vol. 12347, pp. 524–540. Springer, Cham (2020). https://doi.org/10.1007/978-3-030-58536-5_31
24. Rao, D., Visin, F., Rusu, A., Pascanu, R., Teh, Y.W., Hadsell, R.: Continual unsupervised representation learning. In: NeurIPS, vol. 32 (2019)
25. Sanner, A., González, C., Mukhopadhyay, A.: How reliable are out-of-distribution generalization methods for medical image segmentation? In: Bauckhage, C., Gall, J., Schwing, A. (eds.) DAGM GCPR 2021. LNCS, vol. 13024, pp. 604–617. Springer, Cham (2021). https://doi.org/10.1007/978-3-030-92659-5_39
26. Simpson, A.L., et al.: A large annotated medical image dataset for the development and evaluation of segmentation algorithms. CoRR abs/1902.09063 (2019)
27. Srivastava, S., Yaqub, M., Nandakumar, K., Ge, Z., Mahapatra, D.: Continual domain incremental learning for chest X-Ray classification in low-resource clinical settings. In: Albarqouni, S., et al. (eds.) DART/FAIR 2021. LNCS, vol. 12968, pp. 226–238. Springer, Cham (2021). https://doi.org/10.1007/978-3-030-87722-4_21

28. Venkataramani, R., Ravishankar, H., Anamandra, S.: Towards continuous domain adaptation for medical imaging. In: IEEE 16th ISBI, pp. 443–446. IEEE (2019)
29. Vokinger, K.N., Gasser, U.: Regulating AI in medicine in the United States and Europe. Nat. Mach. Intell. **3**(9), 738–739 (2021)
30. Zhang, J., Gu, R., Wang, G., Gu, L.: Comprehensive importance-based selective regularization for continual segmentation across multiple sites. In: de Bruijne, M., et al. (eds.) MICCAI 2021. LNCS, vol. 12901, pp. 389–399. Springer, Cham (2021). https://doi.org/10.1007/978-3-030-87193-2_37
31. Zheng, E., Yu, Q., Li, R., Shi, P., Haake, A.: A continual learning framework for uncertainty-aware interactive image segmentation. In: Proceedings of the AAAI Conference on Artificial Intelligence, vol. 35, pp. 6030–6038 (2021)

Adaptive Optimization with Fewer Epochs Improves Across-Scanner Generalization of U-Net Based Medical Image Segmentation

Rasha Sheikh(✉), Morris Klasen, and Thomas Schultz

University of Bonn, Bonn, Germany
{rasha,klasen,schultz}@cs.uni-bonn.de

Abstract. The U-Net architecture is widely used for medical image segmentation. However, accuracy has been observed to drop, sometimes dramatically, when U-Nets are trained on images that have been acquired with a specific scanner, and are applied to images from another scanner. This indicates an overfitting to image characteristics that are irrelevant to the semantic contents, and is usually mitigated with data augmentation. We argue that early stopping additionally improves across-scanner generalization, while greatly reducing training times. For this, we first observe that the widely used stochastic gradient descent (SGD) trains different U-Net layers at different speeds, and demonstrate that this problem is reduced by switching to AvaGrad, a recently proposed adaptive optimizer. On two different datasets, this allows us to match accuracies from nnUNets with default settings, 1000 epochs of SGD, by training for only 50 epochs with AvaGrad, and to exceed their results in the across-scanner setting. This benefit is specific to combining adaptive optimization and early stopping, since it can be matched neither by SGD with a low number of epochs, nor by Avagrad with many epochs. Finally, we demonstrate that the choice of optimizer can have important implications for domain adaptation. In particular, the SpotTUnet, which was recently proposed to automatically select layers for fine-tuning, arrives at very different policies depending on the optimizer.

1 Introduction

The U-Net architecture [9] has achieved state-of-the-art results for many different medical image segmentation tasks. However, its ability to generalize to images that have been acquired with a different scanner than the images it was trained on is often limited. In practice, generalization depends on many factors, including the network's architecture, how the input data is pre-processed, and the learning scheme. Our work investigates two of these factors, the type of

---

Supplementary Information The online version contains supplementary material available at https://doi.org/10.1007/978-3-031-16852-9_12.

K. Kamnitsas et al. (Eds.): DART 2022, LNCS 13542, pp. 119–128, 2022.
https://doi.org/10.1007/978-3-031-16852-9_12

optimizer and the stopping time. We suggest that widely used settings for them are not optimal with respect to across-scanner generalization, and that this has important implications for domain adaptation techniques.

The nnU-Net framework [5] automatically configures itself for different segmentation tasks [4,7]. Some of the design choices in the framework are adapted to the dataset that the model is trained on, whereas others were found to work well in different challenges and are fixed. These include having SGD as the optimizer with a learning rate of 0.01, and training for a long time, namely, 1000 epochs. Our work is based on the intuition that stopping earlier, which is widely used as a way of regularizing deep networks, should reduce overfitting and therefore improve across-scanner generalization. However, we found that SGD with few epochs does not sufficiently train all U-Net layers, due to a very different effective speed at which different layers are trained. To mitigate this problem, we investigate adaptive optimization.

Adam and SGD are two of the most commonly used methods in the optimization of a network's parameters during training. For vision tasks with convolutional networks [2,5], non-adaptive optimizers, such as SGD, are frequently used with the belief that they generalize better to unseen data [13]. However, recent work [3,10] has shown that adaptive methods can actually outperform non-adaptive ones if they are properly tuned. AvaGrad [10] is a recently published adaptive optimizer. It is similar in principle to Adam, where running averages of the gradients and their squared values are used to update the network's weights. However, it is different in that it decouples the effective learning rate and the adaptability parameter ϵ. If ϵ is large enough, the only parameter left to tune is the learning rate, which makes the cost of the hyperparameter search similar to that with SGD.

In our experiments, we replace SGD with AvaGrad, fix the learning rate and ϵ, and train for a relatively short number of epochs. We test the performance of the trained models on two datasets that exhibit a domain shift between source and target domains, and observe an improvement in performance on both of them. We also show that the choice of optimizer has a great influence on the fine-tuning policies learned by Zakazov et al. [14].

2 Materials and Methods

2.1 Datasets

We selected two public datasets for our experiments, the Calgary-Campinas-359 (CC-359) dataset [12] and the Multi-Centre, Multi-Vendor and Multi-Disease Cardiac Image Segmentation (M&Ms) dataset [1]. They both provide MR images that have been acquired with different scanners, which we can use as the separate domains in our experiments. The **CC-359** dataset consists of 359 3D brain MR images collected from sites with scanners that differ in their type and field strength, resulting in 6 domains: GE 15, GE 3, Philips 15, Philips 3, Siemens 15, and Siemens 3. We use GE 3 as our source domain and the others as the target domains. The groundtruth for this dataset corresponds to skull-stripping segmentation masks. The **M&Ms** dataset consists of cardiac MR images from 345 subjects. These were collected from sites with different scanners: Siemens (A), Philips (B), GE (C), and Canon (D). We use domain B as our source domain and the others as the target domains. The segmented regions in this dataset are the left ventricle cavity (LV), the right ventricle cavity (RV), and the left ventricle myocardium (MYO).

2.2 Segmentation Framework

We use the well-known and established nnUNet [5] as our segmentation pipeline. Its backbone is based on the U-Net [9] architecture with an encoder and decoder-like structure and skip connections in between. The depth of the network. i.e. number of down/upsampling operations is adaptive and depends on the dataset median image size. As a preprocessing step, the images are normalized to follow a standard distribution. Training data is augmented with various transformations including rotation, scaling, Gaussian blurring, and Gamma augmentation among others. The network is trained with the SGD optimizer with a poly learning rate scheduler, and it uses both the dice and cross-entropy as the loss function. In our experiments we focus on the 2D variant of nnUNet. To evaluate the performance, we use the Dice score for the M&Ms dataset, Dice $= 2(\sum \hat{y}y)/(\sum \hat{y}+\sum y)$, where $\hat{y}$ and y are the predicted and true labels respectively. For the CC-359 dataset, we follow [11] and use the surface Dice score [8] instead. This score computes how much of the predicted and ground-truth surfaces overlap within a given distance tolerance. Since the segmentation task for this dataset is skull-stripping, this measure is deemed more informative because of the larger focus on the brain contour.

2.3 AvaGrad Optimizer

Similar to the Adam optimizer, AvaGrad is an adaptive method where the effective learning rate for each parameter (e.g. network weights) is adapted according to the running averages of the corresponding gradients. Differently however, AvaGrad decouples the adaptability parameter ϵ from the learning rate. This is achieved by normalizing the learning rate vectors before using them to update the parameters. In our experiments, for both datasets we use the values 10 for the learning rate and 0.1 for ϵ. These were found using a separate validation set from domain B of the M&Ms dataset. We show below the relevant equations from [10], where w_t and g_t denote the parameter to be updated and its gradient, m_t and v_t denote the running averages of the gradient and the gradient squared, d is the dimension of η_t, and finally α_t and ϵ are the learning rate and the adaptability parameter.

$$m_t = \beta_{1,t} m_{t-1} + (1-\beta_{1,t}) g_t, \qquad \eta_t = \frac{1}{\sqrt{v_{t-1}+\epsilon}},$$

$$w_{t+1} = w_t - \alpha_t \frac{\eta_t}{\|\eta_t/\sqrt{d}\|_2} \odot m_t, \qquad v_t = \beta_{2,t} v_{t-1} + (1-\beta_{2,t}) g_t^2$$

2.4 SpotTUNet

If we have annotated data from target domains, we can fine-tune the source-domain model on each target. Zakazov et al. [14] proposed the SpotTUnet, where an additional ResNet-34 is employed to automatically choose which layers should be fine-tuned. For that, we extract the input features, which get reduced to a 2×32 dimensional output. Each of the 64 output logits is passed through a Gumbel-Softmax [6] which would then be mapped to a probability for a UNet layer to be chosen for fine-tuning or to remain frozen. The original SpotTune also proposed a regularization constraint for the global policy aimed at consolidating the global fine-tuning policy. The SpotTUNet loss with added penalty term is defined as $\mathcal{L} = \mathcal{L}_{segm} + \lambda \sum_{l=1}^{64}(1 - I_l(x))$, where $I_l(x)$ is the binary indicator for the l-th frozen layer based on the image input x. We evaluate the SpotTune performance for 7 and 800 target domain slices available for fine-tuning.

Table 1. Surface Dice performance on the Calgary-Campinas dataset. Largest means in each row are bold, statistical significance between S-1000 and the other columns is indicated with an asterisk.

Domain	S-1000	S-25	S-50	A-25	A-50	AG-25	AG-50	AG-1000
GE15	0.9068	0.8345*	0.8782*	0.8970*	0.9076	0.9152*	**0.9214***	0.8181*
Philips 15	0.9407	0.9102*	0.9330*	0.9421	0.9472*	**0.9484***	0.9480*	0.8009*
Philips 3	0.8685	0.7252*	0.7773*	0.8570	0.8345*	0.8848*	**0.9032***	0.7135*
Siemens 15	0.9176	0.8409*	0.8841*	0.9206	0.9219*	0.9324*	**0.9344***	0.7734*
Siemens 3	0.8191	0.8449*	0.8579*	0.9037*	0.9029*	**0.9061***	0.8932*	0.8398
GE 3 Test	0.9645	0.9594	0.9629	0.9622	0.9646	**0.9650**	0.9641	0.9324*

3 Experiments

3.1 Baseline Performance and Early-Stopping

We show in Tables 1 and 2 the baseline performance with nnUNet in the first column. We refer to this experiment as S-1000 since the default configuration of nnUnet uses the SGD optimizer and trains for 1000 epochs. We next show the performance of the adaptive optimizers, Adam (A) and AvaGrad (AG), with fewer epochs. For comparison, we also train with SGD for the same number of epochs, and with AvaGrad for 1000 epochs. No labels from the target domains were used. On the CC dataset, AvaGrad with few epochs achieves the highest mean scores on all domains. According to a Bonferroni corrected Wilcoxon signed-rank test, all differences between S-1000 and AG-50, which we propose as an improved default, are statistically significant. This improvement is achieved at greatly reduced computational cost, due to a much smaller number of epochs.

Similarly, on the M&Ms dataset (Table 2), we either observe a benefit or a comparable performance from using AvaGrad with few epochs. Qualitative results from both datasets can be found in Fig. 1. The Supplementary Material

Table 2. Volumetric Dice performance on the M&Ms dataset. Largest means in each row are bold, statistical significance between S-1000 and the other columns is indicated with an asterisk.

	Class1 (LV)							
Domain	S-1000	S-25	S-50	A-25	A-50	AG-25	AG-50	AG-1000
A	0.6065	0.6957	0.6663	0.7131	0.7506*	0.7524*	**0.7845***	0.7355*
C	0.8700	0.8660	0.8693	0.8784	0.8758	**0.8862***	0.8754	0.8734
D	0.8805	0.8833	0.8829	0.8767	0.8745	**0.8919**	0.8866	0.8912
B Test	0.8850	0.8977	0.8932	0.9019	0.9024	0.8988	**0.9063**	0.8937
	Class2 (MYO)							
Domain	S-1000	S-25	S-50	A-25	A-50	AG-25	AG-50	AG-1000
A	0.5335	0.5697	0.5468	0.5785	0.6261	0.6200*	**0.6644***	0.6400*
C	0.8064	0.7994	0.8015	0.8091	0.8010	**0.8170**	0.8143	0.8143
D	0.8127	0.8031	0.8084	0.8126	0.8121	0.8112	0.8134	**0.8163**
B Test	0.8463	0.8461	0.8468	0.8569	0.8561	0.8537	**0.8625**	0.8510
	Class3 (RV)							
Domain	S-1000	S-25	S-50	A-25	A-50	AG-25	AG-50	AG-1000
A	0.5000	0.5578	0.5058	0.5821	0.6392*	0.6187	**0.6973***	0.6582*
C	0.8093	0.8196	0.8179	0.8045	0.8241	0.8311	**0.8451**	0.8305
D	0.8028	0.7682*	0.7645*	0.7766	0.8007	0.8027	0.8041	**0.8454**
B Test	0.8635	0.8535	0.8534	0.8535	0.8570	0.8619	**0.8714**	0.8627

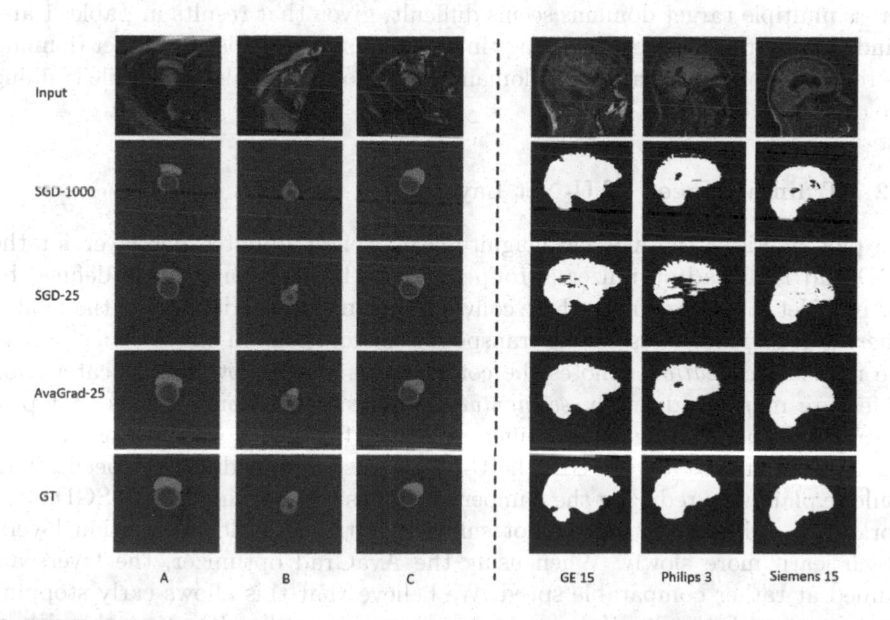

Fig. 1. Qualitative results on the M&Ms dataset on the left (yellow: RV, blue: LV, green: MYO), and on the Calgary-Campinas dataset on the right. (Color figure online)

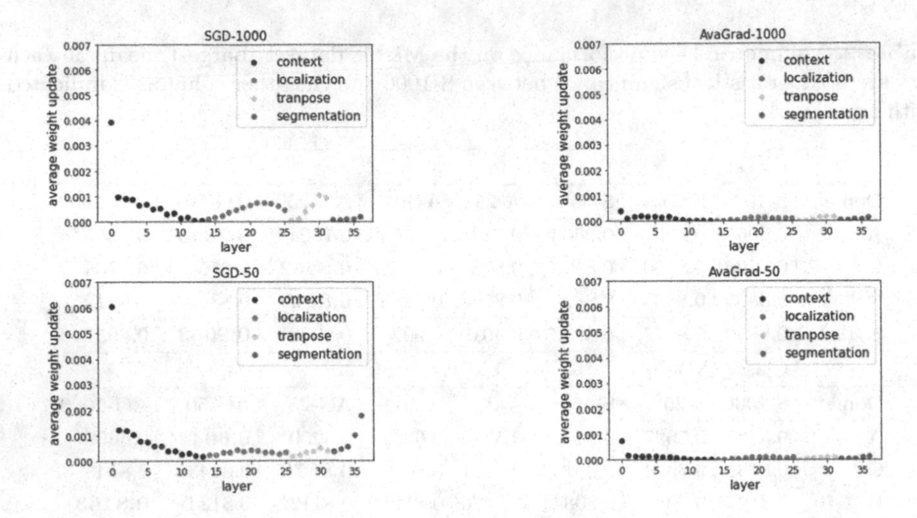

Fig. 2. Average weight update for SGD on the left and AvaGrad on the right.

also includes violin plots that show the distribution of results across the different subjects on both datasets.

We note that, throughout this work, we use the term *early stopping* to simply denote training for much fewer epochs than the nnUNet default. Introducing a mechanism that fine-tunes the number of epochs for optimal generalization across multiple target domains seems difficult, given that results in Tables 1 and 2 indicate that the perfect stopping time depends on the specific target domain. Moreover, in practice, the target domain might not even be known while training the original model.

3.2 Training Speed of U-Net Layers

We plot in Fig. 2 the average magnitude of weight updates per layer for the SGD and AvaGrad optimizers, grouped by the type of convolution defined by the nnUNet. *context* denotes those convolutions in the encoder part of the model, whereas *transpose* refers to the transposed convolutions in the decoder part of the model. *localization* denotes the convolutions that follow the concatenation of feature maps, and finally *segmentation* refers to the convolutions that produce the segmentation maps at different levels of the network. We observe that particularly in the case of SGD, the U-Net layers train at different speeds. This could explain why reducing the number of epochs when training with SGD often works less well, since it might not sufficiently train the low-resolution layers, which learn more slowly. When using the AvaGrad optimizer, the layers are trained at rather comparable speed. We believe that this allows early stopping to reduce overfitting to the source domain, while still mitigating the risk of underfitting individual layers.

3.3 No Augmentation

The nnUNet augments the training data with transformations that simulate plausible differences between scanners. We investigate the relative benefits from augmentation and early stopping through an ablation study that deactivates those augmentations. Table 3 summarizes these results for the Calgary-Campinas dataset. The corresponding table for the M&Ms dataset can be found in the Supplementary Material. Compared to Tables 1 and 2, we observe that across-scanner generalization suffers drastically from this when training for 1000 epochs. This effect is much reduced when training with AvaGrad and early stopping. Best results are obtained when combining both types of regularization.

Table 3. Surface dice performance on the CC dataset when training with no augmentations

Domain	S-1000-noAug	S-25-noAug	A-25-noAug	AG-25-noAug
GE15	0.66380	0.76537	**0.83760**	0.81029
Philips 15	0.85827	0.87022	0.89176	**0.91129**
Philips 3	0.69653	0.62460	0.65254	**0.80440**
Siemens 15	0.80643	0.80820	0.85617	**0.88390**
Siemens 3	0.63098	**0.70149**	0.57215	0.63266
GE 3 Test	0.92995	0.94423	0.94547	0.95422

3.4 SpotTUNet

We evaluate the differences between SGD and AvaGrad under several SpotTUNet settings. For direct comparison, we retain the same architecture and experimental setup as provided by [14][1], where the baseline model was trained for 60 epochs and the spottuned model for 100 epochs. Figure 3 shows that using the surface dice as metric and the SGD as an optimizer, we can spottune the Target Domains (TDs) individually to increase the average target domain performance. Changing the optimizer to AvaGrad immediately achieves an increase of 28.1% surface dice across the TDs. Like with SGD, spottuning with AvaGrad further improves the surface dice. We plot the corresponding SpotTUNet policy visualizations in Figs. 5a and 5b. For SGD, they agree with the finding of Zakazov et al. [14] in that early encoder layers get fine-tuned. This pattern changes with AvaGrad, where fine-tuning focuses on the lowest-resolution layers.

[1] https://github.com/neuro-ml/domain_shift_anatomy.

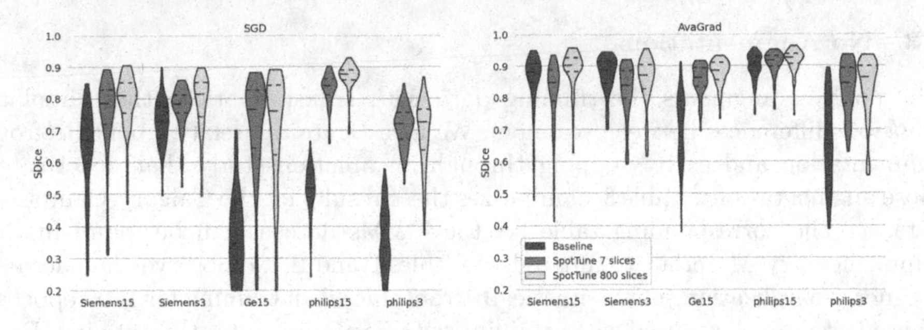

Fig. 3. SpotTUNet performance on the Calgary-Campinas dataset.

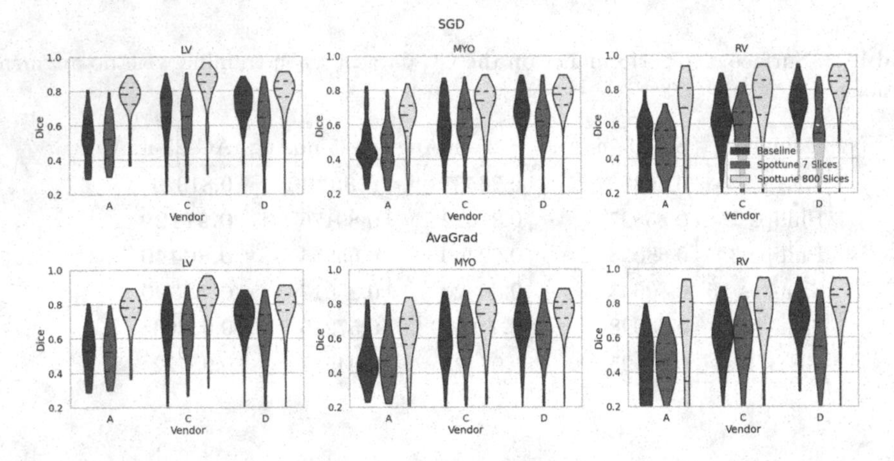

Fig. 4. SpotTUNet performance on the M&Ms dataset.

To spottune on the M&Ms dataset, we use B as the source domain and fine-tune the other domains as targets. Again, changing to AvaGrad increases the baseline performance on the TDs. However, due to the comparatively smaller number of annotated training slices and stronger inhomogeneity between the domains, the models overfit and degenerate for both SGD and AvaGrad using only 7 slices (Fig. 4). That is further highlighted by the policy visualization in Fig. 5c as both SGD and AvaGrad (7 slices) fine-tune layers across the whole model. Refinement with few slices introduces a risk of overfitting. We believe that, with the current Spottuning method, this sometimes outweighs the benefit of introducing domain-specific information. On the other hand, when providing 800 annotated slices for spottuning, we achieve higher average dice scores across all TDs, while AvaGrad still outperforms SGD. The corresponding Tables for both datasets can be found in the Supplementary Material.

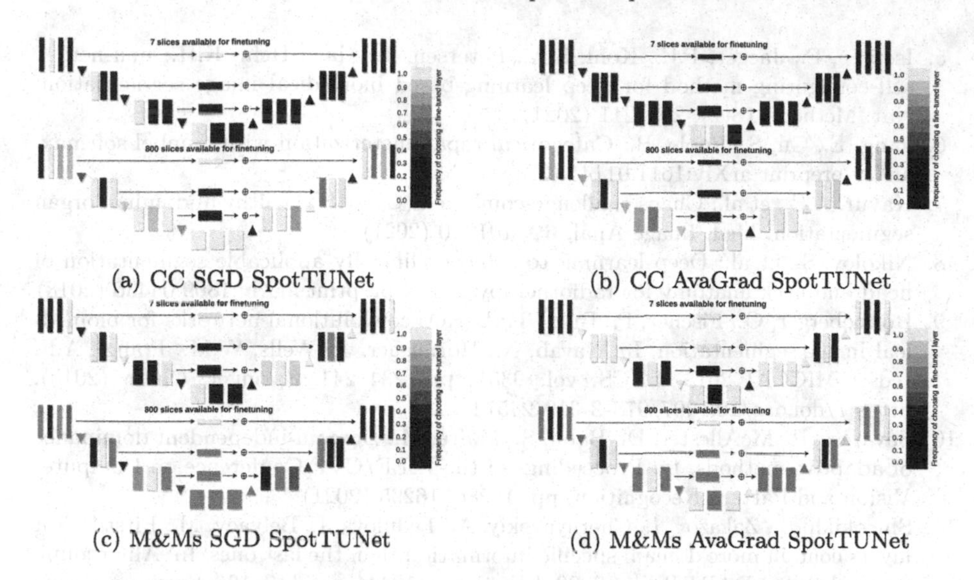

(a) CC SGD SpotTUNet (b) CC AvaGrad SpotTUNet

(c) M&Ms SGD SpotTUNet (d) M&Ms AvaGrad SpotTUNet

Fig. 5. Visualization of the learned SpotTUnet policy for the CC and M&Ms dataset.

4 Conclusion

In this work, we argue that a widely used training strategy for U-Nets, SGD with a large number of epochs, is not optimal with respect to generalization to other domains. In particular, we demonstrate that the use of adaptive optimizers together with early stopping improves generalization across scanners on two different datasets. An additional advantage of early stopping is computational efficiency. This is especially beneficial in cases which involve frequent re-training, such as ensembling or automated configuration. Finally, we show that the choice of optimizer influences the policies that are used to fine-tune models when annotated target data is available.

References

1. Campello, V.M., et al.: Multi-centre, multi-vendor and multi-disease cardiac segmentation: the M&MS challenge. IEEE Trans. Med. Imaging **40**(12), 3543–3554 (2021)
2. Chen, L.C., Zhu, Y., Papandreou, G., Schroff, F., Adam, H.: Encoder-decoder with atrous separable convolution for semantic image segmentation. In: Proceedings of the European Conference on Computer Vision (ECCV), pp. 801–818 (2018)
3. Choi, D., Shallue, C.J., Nado, Z., Lee, J., Maddison, C.J., Dahl, G.E.: On empirical comparisons of optimizers for deep learning. arXiv preprint arXiv:1910.05446 (2019)
4. Heller, N., et al.: The state of the art in kidney and kidney tumor segmentation in contrast-enhanced CT imaging: results of the kits19 challenge. Med. Image Anal. **67**, 101821 (2021)

5. Isensee, F., Jaeger, P.F., Kohl, S.A., Petersen, J., Maier-Hein, K.H.: nnu-net: A self-configuring method for deep learning-based biomedical image segmentation. Nat. Methods **18**(2), 203–211 (2021)
6. Jang, E., Gu, S., Poole, B.: Categorical reparameterization with gumbel-softmax. arXiv preprint arXiv:1611.01144 (2016)
7. Kavur, A.E., et al.: Chaos challenge-combined (CT-MR) healthy abdominal organ segmentation. Med. Image Anal. **69**, 101950 (2021)
8. Nikolov, S., et al.: Deep learning to achieve clinically applicable segmentation of head and neck anatomy for radiotherapy. arXiv preprint arXiv:1809.04430 (2018)
9. Ronneberger, O., Fischer, P., Brox, T.: U-Net: convolutional networks for biomedical image segmentation. In: Navab, N., Hornegger, J., Wells, W.M., Frangi, A.F. (eds.) MICCAI 2015. LNCS, vol. 9351, pp. 234–241. Springer, Cham (2015). https://doi.org/10.1007/978-3-319-24574-4_28
10. Savarese, P., McAllester, D., Babu, S., Maire, M.: Domain-independent dominance of adaptive methods. In: Proceedings of the IEEE/CVF Conference on Computer Vision and Pattern Recognition, pp. 16286–16295 (2021)
11. Shirokikh, B., Zakazov, I., Chernyavskiy, A., Fedulova, I., Belyaev, M.: First U-Net layers contain more domain specific information than the last ones. In: Albarqouni, S., et al. (eds.) DART/DCL -2020. LNCS, vol. 12444, pp. 117–126. Springer, Cham (2020). https://doi.org/10.1007/978-3-030-60548-3_12
12. Souza, R., et al.: An open, multi-vendor, multi-field-strength brain MR dataset and analysis of publicly available skull stripping methods agreement. NeuroImage **170**, 482–494 (2018)
13. Wilson, A.C., Roelofs, R., Stern, M., Srebro, N., Recht, B.: The marginal value of adaptive gradient methods in machine learning. Adv. Neural Inf. Process. Syst. **30** (2017)
14. Zakazov, I., Shirokikh, B., Chernyavskiy, A., Belyaev, M.: Anatomy of domain shift impact on U-Net layers in MRI segmentation. In: de Bruijne, M., et al. (eds.) MICCAI 2021. LNCS, vol. 12903, pp. 211–220. Springer, Cham (2021). https://doi.org/10.1007/978-3-030-87199-4_20

CateNorm: Categorical Normalization for Robust Medical Image Segmentation

Junfei Xiao[1(✉)], Lequan Yu[2], Zongwei Zhou[1], Yutong Bai[1], Lei Xing[3], Alan Yuille[1], and Yuyin Zhou[4]

[1] Johns Hopkins University, Baltimore, USA
xiaojf97@gmail.com
[2] The University of Hong Kong, Pok Fu Lam, Hong Kong
[3] Stanford University, Stanford, USA
[4] UC Santa Cruz, Santa Cruz, USA

Abstract. Batch normalization (BN) uniformly shifts and scales the activations based on the statistics of a batch of images. However, the intensity distribution of the background pixels often dominates the BN statistics because the background accounts for a large proportion of the entire image. This paper focuses on enhancing BN with the intensity distribution of foreground pixels, the one that really matters for image segmentation. We propose a new normalization strategy, named categorical normalization (CateNorm), to normalize the activations according to categorical statistics. The categorical statistics are obtained by dynamically modulating specific regions in an image that belong to the foreground. CateNorm demonstrates both precise and robust segmentation results across five public datasets obtained from different domains, covering complex and variable data distributions. It is attributable to the ability of CateNorm to capture domain-invariant information from multiple domains (institutions) of medical data.

Code is available at https://github.com/lambert-x/CateNorm.

1 Introduction

Normalization techniques are vital to accelerate and stabilize the network training procedure. In addition to batch normalization (BN) [13], alternative techniques have been used, such as group normalization [25], layer normalization [1], and instance normalization [24]. However, these normalization strategies are found to be suboptimal for medical image segmentation [19] because they only estimate the statistics of the images as a whole, which can be easily biased towards the intensity distribution of dominant categories (see Fig. 1).

To address this problem, this paper focuses on enhancing BN with the intensity distribution of foreground pixels—the one that matters for image segmentation. We propose a new normalization strategy, named categorical normalization, named **CateNorm**, to normalize the activations according to categorical statistics. We introduce CateNorm to the U-Net architecture by enforcing the normalization layers to modulate foreground regions (e.g., pancreas, stomach)

K. Kamnitsas et al. (Eds.): DART 2022, LNCS 13542, pp. 129–146, 2022.
https://doi.org/10.1007/978-3-031-16852-9_13

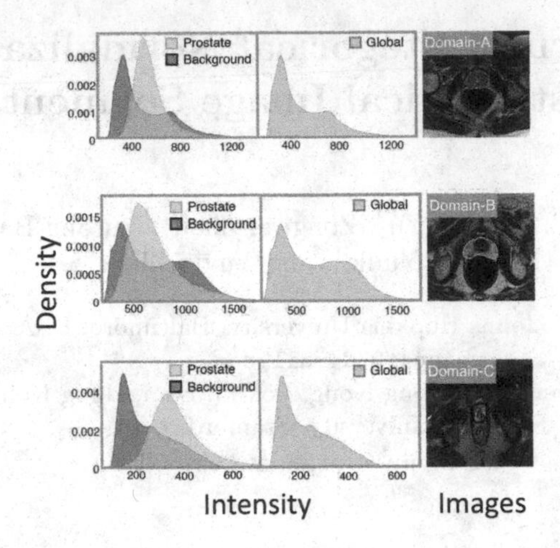

Fig. 1. This paper addresses two limitations in conventional normalization strategies (e.g., BN): (*i*) the intensity distribution of dominant classes, such as background, takes over the global statistics for normalization; (*ii*) the shift of intensity statistics across different domains is substantial (examples in Domains A–C). In response, we propose categorical normalization—instead of normalizing the intensity distribution based on the entire image, CateNorm normalizes the distribution per category.

differently. Specifically, two parallel and complementary schemes are adopted for normalizing the activations—one is the conventional BN to capture the statistics of a batch of images; the other is the proposed CateNorm to integrate the statistics of specific regions with the guidance from the learned categories. As CateNorm conditions on the categorical masks (not available in the inference phase), we hereby introduce a simple yet effective two-stage training strategy: concretely, we first train exclusively using BN for generating the categorical masks and then feed them to CateNorm for updating the semantic-related modulating parameters.

Our extensive experiments on five datasets show that the integration of categorical statistics into normalization strategies can better capture the domain-invariant information from different domain data, thereby robustifying the learned medical representation. Compared with the existing normalization techniques [1,13,24,25], CateNorm achieves more precise and robust segmentation results in various applications, including multi-organ segmentation from CT images and prostate segmentation from MRI images. CateNorm also consistently improves over the previous state of the arts for multi-domain data. This result suggests that CateNorm not only extracts more discriminative features but also compensates for the statistical bias in small distribution shifts and domain gaps.

Our contributions are three-fold: (1) a novel normalization to integrate categorical statistics with BN's general statistics, surpassing existing normalization techniques; (2) a light CateNorm residual block for the U-Net encoder, yielding a prominent performance gain with negligible additional parameters; (3) CateNorm achieves better robustness in complex and variable data distributions.

2 Related Work

Normalization is one of the keys to the success of deep networks. As the most commonly used normalization technique, batch normalization (BN) [13] enables training with larger learning rates and greatly mitigates general gradient issues. Besides, alternative normalization strategies have also been designed for specific scenarios: layer normalization (LN) [1] for recurrent neural networks, instance normalization (IN) [24] for style transfer, group normalization (GN) [25] for small-batch training, etc. Beyond the natural image domain, these normalization strategies have also been successfully applied to medical applications. For example, Kao et al. [15] apply GN for brain tumor segmentation; Isensee et al. [14] apply IN in a self-adaptive framework for various segmentation tasks; Chen et al. [4] apply LN in Transformers for abdominal multi-organ segmentation. A detailed comparison among different normalization strategies for medical semantic segmentation and cross-modality synthesis has been summarized in Zhou et al. [28] and Hu et al. [11]. To offer stronger affine transformation, the latest strategies [7,12] utilize external data to denormalize the features. As class information could be "washed away" with previous strategies, SPADE [20] directly uses class masks to guide the normalization. In addition, for domain adaptation and multi-domain learning, other methods propose to modify BN by modulating [18] or calculating domain-specific [3,19] statistics. Unlike existing strategies, this paper proposes a novel categorical normalization strategy for incorporating different types of statistics.

3 Categorical Normalization (CateNorm)

Proof of Concept: A simple experiment is designed to illustrate how categorical statistics can better mitigate distribution shifts across domains. We align the input image distributions based on the categorical statistics obtained from different datasets. Then we jointly train these heterogeneous datasets on the aligned input images. With class-wise distribution alignment, the model gains an improvement of 0.7% (91.4% vs. 90.7%). This suggests that aligning the data distribution based on class-wise statistics can better mitigate domain shifts between datasets and therefore achieve better results in the joint training setting. This interesting observation further motivates us to design normalization strategies to further leverage local statistics during the learning process.

We devise a two-stage training paradigm. In the first stage, the network is trained with BN to generate the class masks, which are later fed to the second stage to normalize the distribution based on each category. Figure 2 presents the overall pipeline of our approach. In the following, we will first introduce our CateNorm Residual Block (CNRB), following with the overall training and testing pipeline of CateNorm.

CateNorm Residual Block: In addition to BN in residual blocks [10], we introduce categorical normalization scheme while keeping all other components

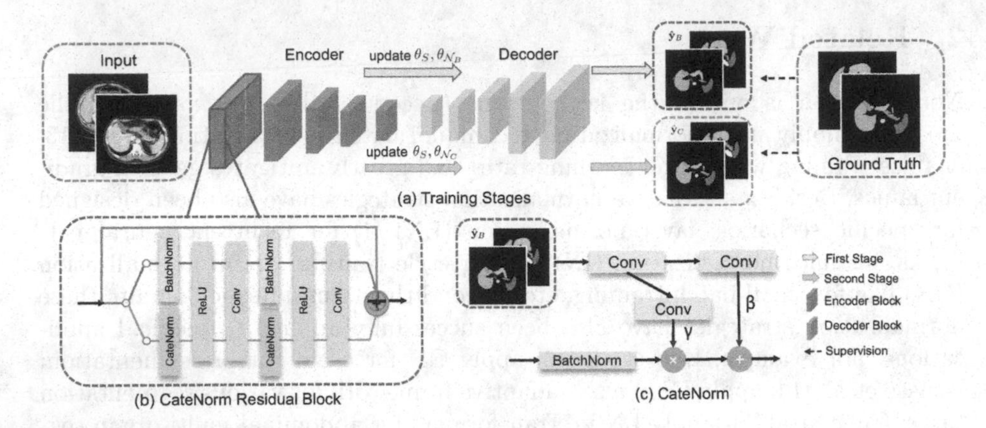

Fig. 2. Framework overview: (a) architecture & training stages of U-Net using BN and CateNorm; (b) design of CateNorm residual block; (c) design of CateNorm.

the same (see Fig. 2(b)). Let x with the spatial resolution of $H \times W$ denotes the input feature map, with a batch of N samples and C number of channels. For the n-th sample at the c-th channel, $x_{n,c,i,j}$ denotes the associated activation value at spatial location (i, j). Then a BN branch and a CateNorm branch are used for fully leveraging the general and categorical statistics, respectively. BN [13] estimates the statistics of a batch of images and then applies affine transformation with learnable parameters γ_c^{BN} and β_c^{BN} at the c-th channel. Specifically, we first compute the mean and standard deviation μ_c and σ_c as follows: $\mu_c = \frac{1}{NHW}\sum_{n,i,j} x_{n,c,i,j}$, $\sigma_c = \sqrt{\frac{1}{NHW}\sum_{n,i,j}\left((x_{n,c,i,j})^2 - (\mu_c)^2\right) + \epsilon}$, where ϵ denotes a small constant for avoiding invalid denominators. And the associated activation map can be then computed as:

$$\gamma_c^{\text{BN}} \cdot \frac{x_{n,c} - \mu_c}{\sigma_c} + \beta_c^{\text{BN}}. \tag{1}$$

CateNorm serves as an addition normalization to integrate categorical statistics. The key difference, compared with BN, is that CateNorm provides a spatially-variant affine transformation which is learned from the corresponding class mask for modulating the activations. Therefore, the modulation parameters $\gamma^{\text{CateNorm}}(\mathbf{m})$ and $\beta^{\text{CateNorm}}(\mathbf{m})$ are no longer C-dimensional vectors as aforementioned, but tensors with spatial resolution $H \times W$. Here $\gamma^{\text{CateNorm}}(\cdot)$ and $\beta^{\text{CateNorm}}(\cdot)$ are learnable functions that act on the given class mask $\mathbf{m}$.

The activations of the CateNorm branch are computed as:

$$\gamma_{c,i,j}^{\text{CateNorm}}(\mathbf{m})\frac{x_{n,c,i,j} - \mu_c}{\sigma_c} + \beta_{c,i,j}^{\text{CateNorm}}(\mathbf{m}). \tag{2}$$

Inspired by SPADE [20], we use a lightweight two-layer convolutional network to learn the modulation functions $\gamma^{\text{CateNorm}}(\cdot)$ and $\beta^{\text{CateNorm}}(\cdot)$, where the first layer is set to output features of $C/2$ number of channels.

Training CateNorm: We inject CateNorm into popular segmentation models, such as U-Net [21] and DeepLabV3+ [5]. The model can be denoted by $\mathcal{F}(\cdot;\theta)$ parameterized by $\theta = \{\theta_s, \theta_{\mathcal{N}_B}, \theta_{\mathcal{N}_C}\}$. $\theta_{\mathcal{N}_B}$ denote the learnable modulating parameters used in the BN branch and $\theta_{\mathcal{N}_C}$ is the learning modulation function that requires the corresponding class mask as an input; the subscripts $\mathcal{N}_B$ and $\mathcal{N}_C$ here denote BN and CateNorm. θ_s stands for all other network parameters.

In the first stage, we exclusive train the model through the BN branch, *i.e.*, only θ_s and $\theta_{\mathcal{N}_B}$ are updated. The goal of the first stage is not only to leverage the global statistics for accelerating training but also to provide class information for learning CateNorm parameters in the second stage. Specifically, the class mask generated from the first stage $\hat{\mathbf{y}}_B$ can be written as:

$$\hat{\mathbf{y}}_B = \mathcal{F}(\theta_S, \theta_{\mathcal{N}_B}; \mathbf{x}). \tag{3}$$

In the second training stage, we train exclusively on the CateNorm branch for exploiting the local statistics, *i.e.*, only update network parameters θ_s and the learnable function $\theta_{\mathcal{N}_C}(\cdot)$. Given the class mask $\hat{\mathbf{y}}_B$, the normalization parameters $\theta_{\mathcal{N}_C}(\hat{\mathbf{y}}_B)$ can be then computed via Eq. (2), and the predicted mask in the second stage $\hat{\mathbf{y}}_C$ can be written as:

$$\hat{\mathbf{y}}_C = \mathcal{F}(\theta_S, \theta_{\mathcal{N}_C}(\hat{\mathbf{y}}_B); \mathbf{x}), \tag{4}$$

where $\hat{\mathbf{y}}_C$ denotes the softmax probability map from the CateNorm branch. With auxiliary class information integrated, this training step aims at enhancing discriminative features, which leads to more accurate and robust segmentation

Overall Training Objective: Given a pair of probability prediction $\hat{\mathbf{y}}$ and the associated ground truth $\mathbf{y} \in \mathbb{L}^D$, the Dice loss and the cross entropy loss are: $\mathcal{L}_{Dice}(\mathbf{y}, \hat{\mathbf{y}}) = \frac{1}{|\mathbb{L}|}\sum_l [1 - \frac{2\sum_{i,j,l} y_{i,j,l}\cdot\hat{y}_{i,j,l}}{\sum_{i,j,l}(y_{i,j,l}^2 + \hat{y}_{i,j,l}^2)}]$, $\mathcal{L}_{CE}(\mathbf{y}, \hat{\mathbf{y}}) = -\frac{1}{D\cdot|\mathbb{L}|}\sum_{i,j,l} y_{i,j,l} \cdot \log(\hat{y}_{i,j,l})$, where $\hat{y}_{i,j,l}$ is the output probability of the l-th class ($l \in \mathbb{L}$) of at spatial location i, j. In our loss function, we use a weighted sum of these two losses, which can be written as: $\mathcal{L}(\mathbf{y}, \hat{\mathbf{y}}) = \lambda\mathcal{L}_{Dice}(\mathbf{y}, \hat{\mathbf{y}}) + (1-\lambda)\mathcal{L}_{CE}(\mathbf{y}, \hat{\mathbf{y}})$, where λ is the balance parameter. Therefore, our overall training objective over these two stages is:

$$\mathcal{L}_{total} = \alpha\mathcal{L}(\mathbf{y}, \hat{\mathbf{y}}_B) + (1-\alpha)\mathcal{L}(\mathbf{y}, \hat{\mathbf{y}}_C), \tag{5}$$

where α is set as 1 in the first stage and 0 in the second stage, for updating $\{\theta_s, \theta_{\mathcal{N}_B}\}$ and $\{\theta_s, \theta_{\mathcal{N}_C}\}$ alternately. In the testing phase, the final prediction is obtained by forwarding twice with Eqs. (3) and (4) sequentially. The whole training procedure is summarized in Appendix Algorithm 1.

4 Experiments

4.1 Dataset and Benchmark

Prostate Segmentation Datasets: Following Liu et al. [19], we use prostate T2-weighted MRI collected from three different domains. (1) 30 samples from Radboud University Nijmegen Medical Centre; (2) 30 samples from Boston Medical Center. Both (1) and (2) are available from NCI-ISBI 2013 challenge (ISBI 13) dataset [2], therefore are denoted as "ISBN-R" and "ISBN-B". (3) 19 samples from Initiative for Collaborative Computer Vision Benchmarking (I2CVB) dataset [17], denoted as "I2CVB". Details of their acquisition protocols are included in Appendix Sect. C.

Table 1. CateNorm vs. three other normalization strategies under the single-domain setting. The performance is measured by Dice score (%). CN denotes our categorical normalization. We report the average performance on five-fold cross-validation, along with the statistic analysis (*$p < 0.5$, **$p < 0.1$, ***$p < 0.05$) between the best and second best methods.

Method	Norm	BTCV	ISBN-R	ISBN-B	I2CVB
Baseline	BN	75.56	88.69	85.20	87.21
Baseline	BN	77.83	89.13	85.99	88.22
Baseline	BN	78.98	90.08	87.22	88.99
Baseline	IN	78.96	89.05	88.37	88.98
Baseline	GN	78.50	89.15	**88.61**	89.16
Ours (block 1)	CN	79.33	91.36	88.55	89.31
Ours (block 1–4)	CN	**80.37****	**91.41****	87.57	**89.79****

Table 2. CateNorm vs. three other methods under the multi-domain setting. CN denotes our categorical normalization. We report the average performance on five-fold cross-validation, along with the statistic analysis (*$p < 0.5$, **$p < 0.1$, ***$p < 0.05$) between ours (bolded) and the best previous method (DSBN [3]).

Method	Norm	BTCV	TCIA	ISBN-R	ISBN-B	I2CVB
Baseline	BN	82.64	87.33	91.50	90.46	90.34
DSBN	BN	82.67	87.83	91.98	90.22	90.10
MS-Net	BN	82.17	87.85	91.93	90.30	89.89
Ours (block 1)	CN	83.28	88.22	92.12	90.76	90.33
Ours (block 1–4)	CN	**83.45***	**88.38****	**92.47*****	**91.17*****	**90.79***

Abdominal Multi-organ Segmentation Datasets: We use abdominal CT images of two different domains. (1) 30 training cases from the Beyond the Cranial Vault (BTCV) dataset [16]; (2) 41 cases from the Cancer Image Archive

(TCIA) Pancreas-CT dataset [6,22,23], where the multi-class annotation can be acquired from Gibson et al. [9]. For single-domain experimental settings, 8 organs (spleen, left kidney, right kidney, gallbladder, pancreas, liver, stomach, and aorta) are evaluated following the settings in Fu et al. [8] and Chen et al. [4]. For multi-domain experiments, right kidney and aorta are excluded and we evaluate the remaining 6 organs which are labeled in both datasets.

4.2 Robust Performance on Single- and Multi-Domain Data

We compare the proposed CateNorm strategy with various normalization methods, including BN [13], IN [24], and GN [25]. In addition, for multi-domain settings, we also compare with state-of-the-art multi-domain learning approaches, including DSBN [3], MS-Net [19]. To ensure a fair comparison, we implement UNet with residual blocks in the encoder [19,27] for all baseline methods.

Single-Domain Results: As shown in Table 1, even with strong data augmentation, our CateNorm still yield a solid performance gain on all four datasets. For instance, on the prostate dataset "ISBN-R" and the multi-organ segmentation dataset, CateNorm outperforms BN by a large margin of 1.33% and 1.39% in average Dice. While different normalization methods (e.g., GN, IN) may behave similarly, our CateNorm consistently achieves better results compared with all other methods. We also compare two different configurations of CateNorm: 1) *CateNorm (block 1)* only replaces the first encoder block with CNRB; 2) *CateNorm (block 1–4)* replaces all of the first four encoder blocks with CNRBs. For prostate segmentation, we find that both variants show a solid improvement while CateNorm *(block 1)* with only few additional parameters performs similarly as CateNorm *(block 1–4)* (i.e., 89.74% vs. 89.59%). On the contrary, for multi-organ segmentation, *block 1* demonstrates inferior results than *block 1–4* (i.e., 79.33% vs. 80.37%). This suggests that CNRB can bring additional benefits for complex tasks such as multi-organ segmentation. For the relatively simpler binary segmentation task, *block 1* might be enough to learn a good model, therefore using more CNRBs does not lead to further performance gain. A detailed study regarding where to add CNRBs has been illustrated in Sect. 4.3.

Multi-domain Segmentation Results: We evaluate our method using the same multi-domain setting as in Liu et al. [19]. To demonstrate the effectiveness of CateNorm, we compare the performance with the baseline and state-of-the-art multi-domain learning methods (i.e., DSBN, MS-Net) on three prostate segmentation datasets and two multi-organ segmentation datasets. As shown in Table 2, under strong data augmentation (e.g., rotation, flipping), DSBN and MS-Net do not yield improvements anymore, while our method still secures a reasonable improvement compared to the baseline. For instance, *block 1–4* outperforms the baseline by 0.81% and 1.05% in average Dice on the BTCV and TCIA dataset, respectively. This indicates that, unlike previous methods which customize the normalization layers for different domains [3,19], our CateNorm can better extract domain-invariant information in the face of a more complex and variable data distribution. Meanwhile, it is also worth mentioning that our approach is complementary to previously domain-specific normalization methods.

Unlike the single-domain setting, *block 1–4* outperforms *block 1* for both prostate segmentation and multi-organ segmentation. We conjecture that this is due to that given a more complex data distribution, more CNRBs can bring additional benefits by imposing semantic guidance on the encoder more densely. More importantly, our CateNorm is flexible to many popular segmentation architectures such as DeepLabV3+ (see Appendix Table 11). Moreover, CateNorm shows great robustness in partially annotated scenarios—detailed studies are provided in Appendix Fig. 3. Qualitative results are in Appendix Fig. 4.

Table 3. Parameter sharing for the two-stage normalization (Dice Score in %)

Method	Sharing	#Params	Dice
U-Net	–	1.0×	90.08
W-Net	✗	2.073×	90.36
Ours(blk1)	✓	1.001×	91.36
Ours(blk1–4)	✓	1.073×	**91.41**

(a) Sharing vs. no-sharing

	Forward	Prostate	Abdominal
Baseline	BN	90.76	84.98
Ours	BN	91.00	85.39
	CN	**91.48**	**85.92**

(b) Mutual benefits of BN and CateNorm.

4.3 Discussion and Ablation Study

Parameter Sharing for the Two-Stage Normalization: To prove the necessity of network sharing except the normalization layers, we also implement our method by using different network parameters θ_s in the two training stages, similar to W-Net [26]. In this implementation, CateNorm and BN are deployed in two independent sub-networks, which are simply concatenated for training and testing. As shown in Table 3a, our method performs much better than W-Net with only about 50% of the parameters. Besides, even only comparing the results in the first stage where only BN is used during inference, as shown in Table 3b, our approach still outperforms the baseline by 0.24% and 0.41% in average Dice. Then in the second stage where CateNorm is used for inference, the performance can be further improved by 0.48% and 0.53%. This indicates that by sharing the rest of the network parameters, the two normalization schemes can mutually benefit each other by leveraging both general and categorical statistics.

Adding CateNorm to the Earlier Encoder: As shown in Sect. 4.2, adding more CateNorm in the U-Net encoder can benefit both prostate and multi-organ segmentation, especially under the multi-domain setting. This observation motivates us to further investigate where to add CateNorm, as this can help us design better configurations of CateNorm which achieve higher performance without incurring much computation cost. In our experiments, U-Net consists of 5 encoder blocks and 5 decoder blocks. By varying the position to add the CateNorm from blocks 1 to 10, we compare the average Dice score on the BTCV dataset. As shown in Appendix Fig. 5, CateNorm blocks are preferred to be set in early blocks (encoder).

Visualizing activations in CateNorm: Appendix Fig. 6 visualizes the learned γ^{CateNorm} and β^{CateNorm} on different channels of the intermediate CateNorm layers during the second forward. With prior class information as guidance, CateNorm can modulate spatially-adaptive parameters. Such spatial-wise modulation can be complementary to the channel-wise modulation accomplished by BN, and derives more discriminative features that benefit segmentation.

5 Conclusion

We have presented a new normalization strategy, named CateNorm, which complementarily enhance the categorical statistics in BN for robust medical image segmentation. Our CateNorm can be used as an add-on to existing segmentation architectures, such as U-Net and DeepLabV3+. Compared with existing normalization strategies, CateNorm consistently achieves superior results, even with complex and variable data distributions. We believe that the proposed normalization strategy could also improve natural image segmentation and plan to explore it in the future work.

Acknowledgments. This work was supported by the Lustgarten Foundation for Pancreatic Cancer Research. We also thank Quande Liu for the discussion.

A Details of Aligning Input Distribution Algorithm

Assume that we have N source domains $S_1, S_2, S_3, ..., S_N$, with $M_1, M_2, M_3, ...,$ M_N examples respectively, where the i-th domain source domain S_i consists of an image set $\{\mathbf{x}_{i,j} \in \mathbb{R}^{D_{i,j}}\}_{j=1,...,M_i}$ as well as their associated annotations. Our goal is to align the image distributions of these source domains with the target domain T based on the class-wise (region-wise) statistics. The algorithm can be illustrated as the following steps:

Step 1: Calculate class-wise statistics of each case
Firstly, we calculate the mean and standard deviation of each case in both the source domain and the target domain.

$$\mu_{i,j}^c = \frac{\sum_{k=1}^{|D_{i,j}^c|} \mathbf{x}_{i,j,k}^c}{|D_{i,j}^c|}, \tag{6}$$

$$\sigma_{i,j}^c = \sqrt{\frac{1}{|D_{i,j}^c|} \sum_{k=1}^{|D_{i,j}^c|} (\mathbf{x}_{i,j,k}^c - \mu_{i,j}^c)^2}, \tag{7}$$

where $\mathbf{x}_{i,j}^c$ denotes the pixels which belong to the c-th class (region) in image $\mathbf{x}_{i,j}$, with the number of pixels denoted as $|D_{i,j}^c|$. As a special case, $i = T$ indicates the target domain.

Step 2: Estimate aligned (new) class-wise statistics
Next, we calculate the mean of the statistics over all examples obtained in each domain as follows:

$$\bar{\mu}_i^c = \frac{\sum_{j=1}^{M_i} \mu_{i,j}^c}{M_i}, \tag{8}$$

$$\bar{\sigma}_i^c = \frac{\sum_{j=1}^{M_i} \sigma_{i,j}^c}{M_i}. \tag{9}$$

Based on the $\bar{\mu}_i^c$, we now estimate the new class-wise mean $\tilde{\mu}_{i,j}$ for each case of the source domain S_i as follows:

$$\tilde{\mu}_{i,j}^c = \frac{\mu_{i,j}^c - \bar{\mu}_i^c}{\sqrt{\frac{\sum_{j=1}^{M_i}(\mu_{i,j}^c - \bar{\mu}_i^c)^2}{M_i}}} \cdot \sqrt{\frac{\sum_{j=1}^{M_T}(\mu_{T,j}^c - \bar{\mu}_T^c)^2}{M_T}} + \bar{\mu}_T^c, \tag{10}$$

where M_T denotes the number of cases in the target domain T. Similarly, the new standard deviation $\tilde{\sigma}_{i,j}$ can be computed by:

$$\tilde{\sigma}_{i,j}^c = \frac{\sigma_{i,j}^c - \bar{\sigma}_i^c}{\sqrt{\frac{\sum_{j=1}^{M_i}(\sigma_{i,j}^c - \bar{\sigma}_i^c)^2}{M_i}}} \cdot \sqrt{\frac{\sum_{j=1}^{M_T}(\sigma_{T,j}^c - \bar{\sigma}_T^c)^2}{M_T}} + \bar{\sigma}_T^c. \tag{11}$$

Step 3: Align each case with the estimated statistics
Based on the computed new mean and standard deviation $\tilde{\mu}_{i,j}$, $\tilde{\sigma}_{i,j}$, the aligned image $\tilde{\mathbf{x}}_{i,j}$ can be computed as:

$$\tilde{\mathbf{x}}_{i,j}^c = \frac{\mathbf{x}_{i,j}^c - \mu_{i,j}^c}{\sigma_{i,j}^c} \cdot \tilde{\sigma}_{i,j}^c + \tilde{\mu}_{i,j}^c. \tag{12}$$

B Implementation Details

(See Tables 4, 5).

Table 4. Data preprocessing.

Step	Prostate	Abdominal
1	Center-cropping	Window range clipping [−125, 275]
2	Out-of-mask slice cropping	Out-of-mask slice cropping
3	Resizing	Resizing
4	Z-score normalization	Z-score normalization

Table 5. Experimental setting.

Config	Value
Training iterations	9000
Optimizer	Adam
Initial learning rate	1e−3
Optimizer momentum	$\beta_1, \beta_2 = 0.9, 0.999$
Batch size	4 (single) 6 (multi)
Learning rate schedule	Plateau scheduler
Dice/CE balance factor λ	0.5 (abdominal) 1.0 (prostate)
Augmentation	Horizontal flipping (prostate only) + random rotation
Validation strategy	5-fold
Evaluation metric	Dice Score (%) and ASD (mm)

C Details of the prostate datasets

(See Tables 6).

Table 6. Details of the 3 prostate segmentation datasets.

Dataset	# Cases	Field strength (T)	Resolution (in/through plane)	Manufacturer
ISBN-R	30	3	0.6–0.625/3.6–4	Siemens
ISBN-B	30	1.5	0.4/3	Philips
I2CVB	19	3	0.67–0.79/1.25	Siemens

D Training procedure of CateNorm

Algorithm 1. Training procedure of CateNorm

Require: Images and labels $\mathbf{x}$, $\mathbf{y}$;
Network parameters $\theta = \{\theta_S, \theta_{\mathcal{N}_B}, \theta_{\mathcal{N}_C}\}$;
Training iterations τ;
Ensure: Optimized parameters θ_S, $\theta_{\mathcal{N}_B}$, $\theta_{\mathcal{N}_C}$;
1: t $\leftarrow$ 0;
2: Initialize θ_S, $\theta_{\mathcal{N}_B}$ with the pretrained model and randomly initialize $\theta_{\mathcal{N}_C}$;
3: **while** $t < \tau$ **do**
4: Compute the class mask $\hat{\mathbf{y}}_B$;
5: $\alpha \leftarrow 1$;
6: Update $\theta_s, \theta_{\mathcal{N}_B} \leftarrow \min_{\theta_s, \theta_{\mathcal{N}_B}} \mathcal{L}_{total}$;
7: Detach $\hat{\mathbf{y}}_B$ from gradient calculation;
8: Compute the class mask $\hat{\mathbf{y}}_C$;
9: $\alpha \leftarrow 0$;
10: Update $\theta_s, \theta_{\mathcal{N}_C}(\cdot) \leftarrow \min_{\theta_s, \theta_{\mathcal{N}_C}} \mathcal{L}_{total}$;
11: $t \leftarrow t + 1$;
12: **end while**

E Average Surface Distance (ASD) Comparison

The detailed average surface distance results of both prostate segmentation and abdominal segmentation tasks can be found in Tables 7 and 8. The proposed CateNorm achieves the lowest average ASD on both tasks, even under the more challenging multi-domain setting (Tables 9, 10, 12).

Table 7. ASD comparison on the abdominal datasets under the multi-domain setting (in mm). Compared with the baseline and other competitive methods, the proposed CateNorm achieves the lowest average ASD.

Method	Forward	BTCV	TCIA	AVG	Spleen	Kid.(L)	Gall	Liver	Stom	Panc
Baseline	BN	1.28	1.17	1.22	0.59	0.59	2.36	0.77	1.93	1.10
DSBN [3]	BN	1.86	**0.90**	1.38	0.51	0.79	3.07	0.76	1.96	1.19
MS-Net [19]	BN	1.61	1.02	1.31	0.52	0.75	2.91	0.91	**1.58**	1.21
Ours (block1)	CN	**1.22**	1.10	**1.16**	0.54	0.58	**2.22**	**0.74**	1.76	1.10
Ours (block1–4)	CN	1.64	0.97	1.30	**0.51**	**0.55**	3.25	0.75	1.75	**1.01**

Table 8. ASD comparison on prostate segmentation datasets under the multi-domain setting (in mm). Compared with the baseline and other competitive methods, the proposed CateNorm achieves the lowest average ASD.

Method	Norm	ISBN-R	ISBN-B	I2CVB	AVG
Baseline	BN	0.64	0.71	1.22	0.86
DSBN [3]	BN	0.56	0.69	1.17	0.81
MS-Net [19]	BN	0.58	0.70	1.32	0.87
Ours (block1)	CN	0.63	0.66	1.27	0.85
Ours (block1–4)	CN	**0.54**	**0.64**	**1.13**	**0.77**

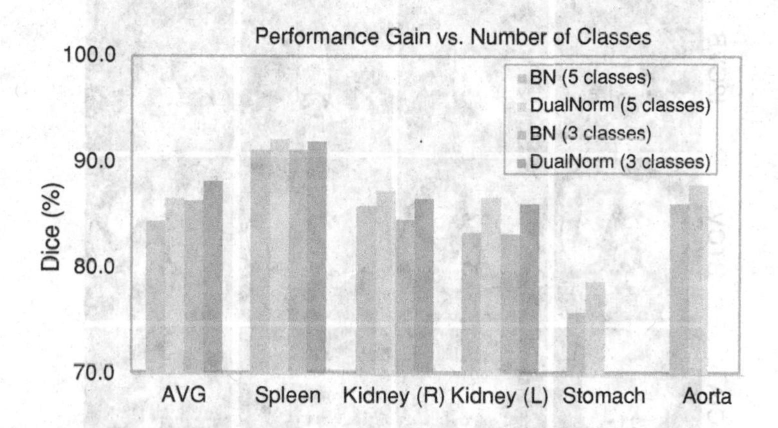

Fig. 3. Performance gain under partial annotation. We compare our method to the baseline with fewer annotated classes (i.e., 3/5). We can see that by partitioning the images into different number of regions, CateNorm consistently achieves better results than BN for all tested organs. This suggests that our algorithm is not sensitive to the number of regions.

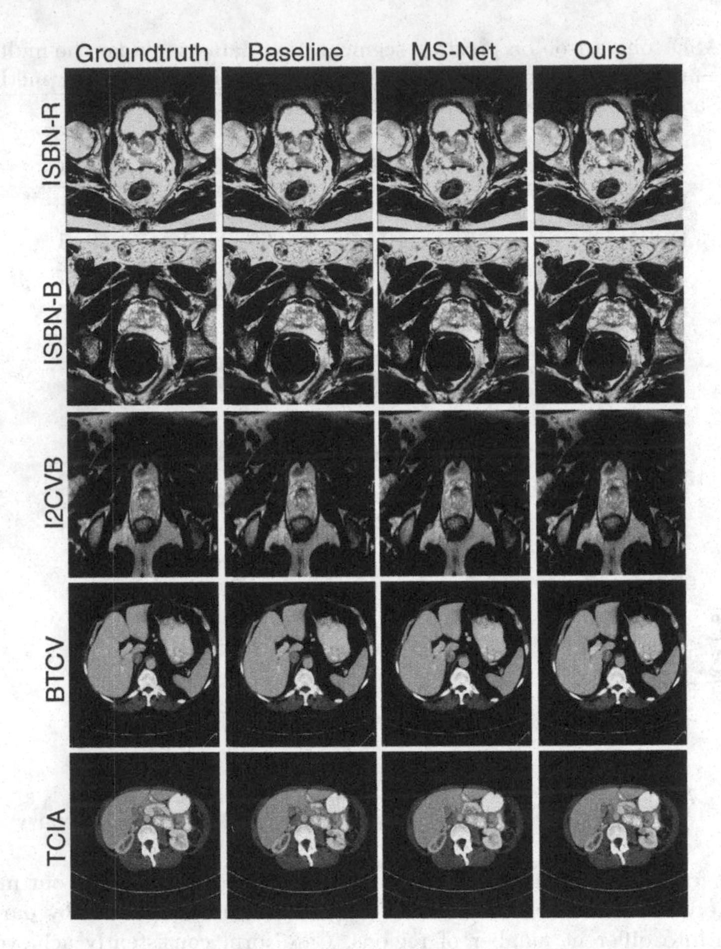

Fig. 4. Qualitative results comparison. We compare our baseline and the other SOTA method under the multi-domain setting on prostate segmentation and abdominal multi-organ segmentation. Results in the first three rows clearly show that our method outperforms others as their results are cracked and incomplete with these unapparent prostate boundaries. And the results in the last two rows show our methods could better suppress inconsistent class information inside a close segmented area (e.g., reducing false positives inside the stomach) and predict hard organs like the pancreas more accurately by incorporating general and categorical statistics.

Table 9. Comparison on the multi-organ segmentation dataset (BTCV) with single-domain setting (Dice Score in %).

Method	Pretrained	Aug	Norm	AVG	Spleen	Kidney (R)	Kidney (L)	Gallbladder	Pancreas	Liver	Stomach	Aorta
Baseline	✗	✗	BN	75.56	92.00	81.92	84.45	50.76	48.89	95.07	68.25	83.16
Baseline	✗	✓	BN	77.83	91.86	84.36	86.99	53.92	51.18	95.11	75.24	83.97
Baseline	✓	✓	BN	78.98	93.43	84.68	87.98	54.54	55.14	95.35	76.19	84.49
Baseline	✓	✓	IN	78.96	91.93	83.70	87.83	54.29	55.08	95.23	78.72	84.93
Baseline	✓	✓	GN	78.50	89.84	85.64	86.26	55.14	55.70	94.71	75.98	84.71
Ours(blk1)	✓	✓	CN	79.33	93.07	85.66	88.26	52.79	54.34	**95.64**	77.38	**87.49**
Ours(blk-4)	✓	✓	CN	**80.37**	**94.63**	**86.29**	**88.64**	**55.51**	**55.91**	**95.64**	**79.80**	86.52

Table 10. Organ-wise results on the multi-organ segmentation datasets under the multi-domain setting (Dice Score in %).

Method	Norm	BTCV	TCIA	AVG	Spleen	Kidney (L)	Gallbladder	Liver	Stomach	Pancreas
Baseline	BN	82.64	87.33	84.98	95.22	93.54	69.63	96.01	85.52	69.97
DSBN	BN	82.67	87.83	85.25	95.42	93.49	69.98	**96.16**	86.11	70.34
MS-Net	BN	82.17	87.85	85.01	95.36	93.10	68.38	95.75	86.77	70.69
Ours (block1)	CateNorm	83.28	88.22	85.75	95.45	**93.63**	70.79	96.13	87.18	71.32
Ours (block1–4)	CateNorm	**83.45**	**88.38**	**85.92**	**95.55**	93.48	**71.53**	96.13	**87.20**	**71.60**

Table 11. CateNorm is compatible to other segmentation models. This table compares performance on multi-domain multi-organ and prostate segmentation with DeepLabv3+ [5] architecture. Our CateNorm consistently outperforms BN.

Backbone	Norm	AVG	BTCV	TCIA	
DeepLabV3+	BN	84.58	81.33	87.83	
DeepLabV3+	CN	**85.42**	**82.13**	**88.72**	
		AVG	ISBN-R	ISBN-B	I2CVB
DeepLabV3+	BN	88.54	90.73	89.35	85.55
DeepLabV3+	CN	**89.21**	**91.26**	**89.78**	**86.59**

Table 12. CateNorm is not sensitive to the warmup length. This table reports average accuracy (%) of our CateNorm under deteriorated pretrained models with fewer pretraining iterations. We reduce the warmup iterations to 450, 1440, and 9000 for multi-domain prostate segmentation experiments, to investigate how our CateNorm performs when warmuped with less iterations.

Warmup Iters	450	1440	9000
Warmup Acc.	73.91%	82.32%	90.10%
Ours Acc.	**91.16%**	**91.17%**	**91.48%**

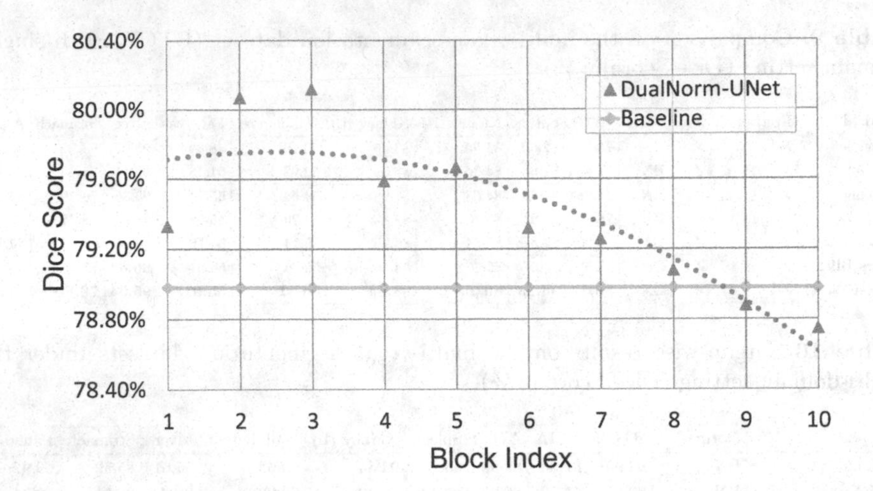

Fig. 5. Set CateNorm block(s) early. This table compares performance with single CateNorm block set in different positions. Adding the CateNorm to the encoder (block index 1–5) always yields better performance than adding to the decoder (block index 6–10). In general, the performance decreases as the block index increases. We believe that it is because the earlier layers in the encoder extract lower-level features that are less discriminative than the decoder features.

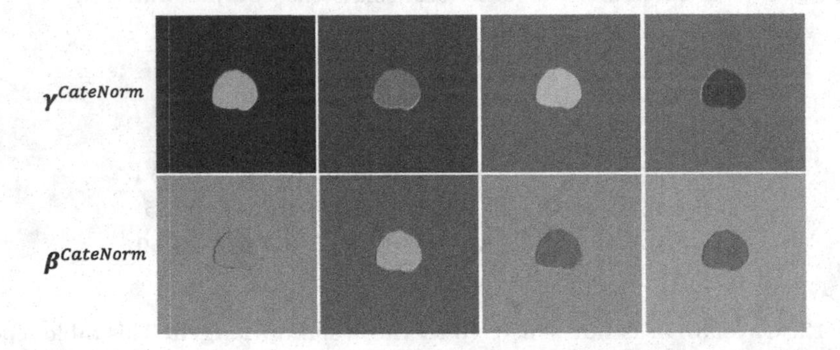

Fig. 6. CateNorm does normalize with semantic information. This figure visualizes the learned γ^{CateNorm} (1st row) and β^{CateNorm} (2nd row) of a CateNorm layer in a CateNorm block on different channels of the intermediate CateNorm layer during the second forward. With prior class information as guidance, CateNorm can modulate spatially-adaptive parameters. Such spatial-wise modulation can be complementary to the channel-wise modulation accomplished by BN, and derives more discriminative features that benefit segmentation.

References

1. Ba, J.L., Kiros, J.R., Hinton, G.E.: Layer normalization (2016)
2. Bloch, N., et al.: NCI-ISBI 2013 challenge: automated segmentation of prostate structures. The Cancer Imaging Archive (2015). http://doi.org/10.7937/K9/TCIA.2015.zF0vlOPv

3. Chang, W.G., You, T., Seo, S., Kwak, S., Han, B.: Domain-specific batch normalization for unsupervised domain adaptation. In: Proceedings of the IEEE/CVF Conference on Computer Vision and Pattern Recognition, pp. 7354–7362 (2019)
4. Chen, J., et al.: TransUNet: transformers make strong encoders for medical image segmentation (2021)
5. Chen, L.-C., Zhu, Y., Papandreou, G., Schroff, F., Adam, H.: Encoder-decoder with atrous separable convolution for semantic image segmentation. In: Ferrari, V., Hebert, M., Sminchisescu, C., Weiss, Y. (eds.) ECCV 2018. LNCS, vol. 11211, pp. 833–851. Springer, Cham (2018). https://doi.org/10.1007/978-3-030-01234-2_49
6. Clark, K., et al.: The cancer imaging archive (TCIA): maintaining and operating a public information repository. J. Digit. Imaging **26**(6), 1045–1057 (2013)
7. Dumoulin, V., Shlens, J., Kudlur, M.: A learned representation for artistic style. In: 5th International Conference on Learning Representations, ICLR 2017, Toulon, France, 24–26 April 2017, Conference Track Proceedings. OpenReview.net (2017). https://openreview.net/forum?id=BJO-BuT1g
8. Fu, S., et al.: Domain adaptive relational reasoning for 3D multi-organ segmentation. In: Martel, A.L., et al. (eds.) MICCAI 2020. LNCS, vol. 12261, pp. 656–666. Springer, Cham (2020). https://doi.org/10.1007/978-3-030-59710-8_64
9. Gibson, E., et al.: Multi-organ abdominal CT reference standard segmentations, February 2018. https://doi.org/10.5281/zenodo.1169361
10. He, K., Zhang, X., Ren, S., Sun, J.: Deep residual learning for image recognition. In: Proceedings of the IEEE Conference on Computer Vision and Pattern Recognition, pp. 770–778 (2016)
11. Hu, S., Yuan, J., Wang, S.: Cross-modality synthesis from MRI to pet using adversarial U-Net with different normalization. In: 2019 International Conference on Medical Imaging Physics and Engineering (ICMIPE), pp. 1–5. IEEE (2019)
12. Huang, X., Belongie, S.: Arbitrary style transfer in real-time with adaptive instance normalization. In: Proceedings of the IEEE International Conference on Computer Vision, pp. 1501–1510 (2017)
13. Ioffe, S., Szegedy, C.: Batch normalization: accelerating deep network training by reducing internal covariate shift. In: International Conference on Machine Learning, pp. 448–456. PMLR (2015)
14. Isensee, F., Kickingereder, P., Wick, W., Bendszus, M., Maier-Hein, K.H.: No new-net. In: Crimi, A., Bakas, S., Kuijf, H., Keyvan, F., Reyes, M., van Walsum, T. (eds.) BrainLes 2018. LNCS, vol. 11384, pp. 234–244. Springer, Cham (2019). https://doi.org/10.1007/978-3-030-11726-9_21
15. Kao, P.-Y., Ngo, T., Zhang, A., Chen, J.W., Manjunath, B.S.: Brain tumor segmentation and tractographic feature extraction from structural MR images for overall survival prediction. In: Crimi, A., Bakas, S., Kuijf, H., Keyvan, F., Reyes, M., van Walsum, T. (eds.) BrainLes 2018. LNCS, vol. 11384, pp. 128–141. Springer, Cham (2019). https://doi.org/10.1007/978-3-030-11726-9_12
16. Landman, B., Xu, Z., Igelsias, J., Styner, M., Langerak, T., Klein, A.: 2015 MICCAI multi-atlas labeling beyond the cranial vault workshop and challenge (2015). https://doi.org/10.7303/syn3193805
17. Lemaître, G., Martí, R., Freixenet, J., Vilanova, J.C., Walker, P.M., Meriaudeau, F.: Computer-aided detection and diagnosis for prostate cancer based on mono and multi-parametric MRI: a review. Comput. Biol. Med. **60**, 8–31 (2015)
18. Li, Y., Wang, N., Shi, J., Liu, J., Hou, X.: Revisiting batch normalization for practical domain adaptation (2016)

19. Liu, Q., Dou, Q., Yu, L., Heng, P.A.: MS-Net: multi-site network for improving prostate segmentation with heterogeneous MRI data. IEEE Trans. Med. Imaging **39**(9), 2713–2724 (2020)
20. Park, T., Liu, M.Y., Wang, T.C., Zhu, J.Y.: Semantic image synthesis with spatially-adaptive normalization. In: Proceedings of the IEEE/CVF Conference on Computer Vision and Pattern Recognition, pp. 2337–2346 (2019)
21. Ronneberger, O., Fischer, P., Brox, T.: U-Net: convolutional networks for biomedical image segmentation. In: Navab, N., Hornegger, J., Wells, W.M., Frangi, A.F. (eds.) MICCAI 2015. LNCS, vol. 9351, pp. 234–241. Springer, Cham (2015). https://doi.org/10.1007/978-3-319-24574-4_28
22. Roth, H., Farag, A., Turkbey, E.B., Lu, L., Liu, J., Summers, R.M.: Data from Pancreas-CT (2016). https://doi.org/10.7937/K9/TCIA.2016.TNB1KQBU, The Cancer Imaging Archive. https://doi.org/10.7937/K9/TCIA.2016.tNB1kqBU
23. Roth, H.R., et al.: DeepOrgan: multi-level deep convolutional networks for automated pancreas segmentation. In: Navab, N., Hornegger, J., Wells, W.M., Frangi, A.F. (eds.) MICCAI 2015. LNCS, vol. 9349, pp. 556–564. Springer, Cham (2015). https://doi.org/10.1007/978-3-319-24553-9_68
24. Ulyanov, D., Vedaldi, A., Lempitsky, V.: Instance normalization: the missing ingredient for fast stylization (2017)
25. Wu, Y., He, K.: Group normalization. In: Ferrari, V., Hebert, M., Sminchisescu, C., Weiss, Y. (eds.) ECCV 2018. LNCS, vol. 11217, pp. 3–19. Springer, Cham (2018). https://doi.org/10.1007/978-3-030-01261-8_1
26. Xia, X., Kulis, B.: W-Net: a deep model for fully unsupervised image segmentation (2017)
27. Yu, L., Yang, X., Chen, H., Qin, J., Heng, P.A.: Volumetric convnets with mixed residual connections for automated prostate segmentation from 3D MR images. In: Thirty-First AAAI Conference on Artificial Intelligence (2017)
28. Zhou, X.Y., Yang, G.Z.: Normalization in training U-Net for 2-D biomedical semantic segmentation. IEEE Robot. Autom. Lett. **4**(2), 1792–1799 (2019)

Author Index

Printed in the United States
by Baker & Taylor Publisher Services

Printed in the United States
by Baker & Taylor Publisher Services